METHODS IN MOLECULAR BIOLOGY

For further volumes:
http://www.springer.com/series/7651

For over 35 years, biological scientists have come to rely on the research protocols and methodologies in the critically acclaimed *Methods in Molecular Biology* series. The series was the first to introduce the step-by-step protocols approach that has become the standard in all biomedical protocol publishing. Each protocol is provided in readily-reproducible step-by-step fashion, opening with an introductory overview, a list of the materials and reagents needed to complete the experiment, and followed by a detailed procedure that is supported with a helpful notes section offering tips and tricks of the trade as well as troubleshooting advice. These hallmark features were introduced by series editor Dr. John Walker and constitute the key ingredient in each and every volume of the *Methods in Molecular Biology* series. Tested and trusted, comprehensive and reliable, all protocols from the series are indexed in PubMed.

ELISA

Methods and Protocols

Edited by

Robert S. Matson

QuantiScientifics LLC, Orange, CA, USA

Humana Press

Editor
Robert S. Matson
QuantiScientifics LLC
Orange, CA, USA

ISSN 1064-3745 ISSN 1940-6029 (electronic)
Methods in Molecular Biology
ISBN 978-1-0716-2902-4 ISBN 978-1-0716-2903-1 (eBook)
https://doi.org/10.1007/978-1-0716-2903-1

This Humana imprint is published by the registered company Springer Science+Business Media, LLC, part of Springer Nature.
The registered company address is: 1 New York Plaza, New York, NY 10004, U.S.A.

Dedication

for

Lielle and Ayla

Preface

This edition of *ELISA: Methods and Protocols* was written during the challenging times associated with the onslaught of the Covid-19 worldwide pandemic. It is truly remarkable to have witnessed the rapid emergence of the recombinant (mRNA/DNA) vaccine strategy to combat this formidable enemy as well as others we have yet to know. We must also acknowledge the tremendous effort of the diagnostic industry to provide reliable testing on such a scale and with very short notice. In the backdrop, there was an army of skilled scientists supporting these efforts. It is very likely that many used the ELISA in support of their important work.

While the enzyme-linked immunosorbent assay (ELISA) is one of the most widely used techniques, its nuances are not always fully understood. Case in point, the revelation that excessive biotin consumption may led to assay interference when using the well-established streptavidin-biotin reporter system. You will find that the authors offer great insight into the theory and practice of the ELISA with useful notes to help you through the step-by-step protocols while steering you clear of common pitfalls.

In this compendium, with the introductory chapter we start with a review of the basic formats used in the ELISA, their application, strengths, and weaknesses. This is followed by examining the essential components and features comprising the technique-enzyme, antibody, and sorbent. Approaches to the immobilization of antibodies are provided for those who may find a need to prepare their own immunosorbent. Since the antibody specificity and sensitivity are most critical to the success of the assay, a chapter is dedicated to screening and selection of antibodies. Next, we discuss means to effectively block the sorbent surface to reduce non-specific binding. In another chapter, the advances in surface modification of the plastics commonly used as the sorbent are described. A chapter on interference completes the first part of the book.

In the second half of the book, we highlight applications of the ELISA. These include the determination of cytokines in tissues and cerebral spinal fluid by sandwich ELISA, a microarray-based method for carbohydrate characterization, and a chapter on mycotoxin extraction with determination by competitive immunoassay. Automation of ELISA for high-throughput diagnostic applications, multiplex allergen testing, and adaptation of ELISA for lateral flow are described in other chapters. There is exciting work on a homogeneous immunoassay and a final chapter examining the role of ELISA in biosensors. It is the goal of this book to provide the reader with a basic understanding of the technique with methods and protocols that enable success at performing your enzyme-linked immunosorbent assay.

Orange, CA, USA *Robert S. Matson*

Contents

Contributors

JULIANO ALVES • *R&D Department, Promega Corporation, Madison, WI, USA*
BYOUNGHOON (BRIAN) HWANG • *R&D Department, Promega Corporation, Madison, WI, USA; Abbvie, North Chicago, IL, USA*
LAURIE ENGEL • *R&D Department, Promega Corporation, Madison, WI, USA*
JOHAN ENGSTRÖM • *Gyros Protein Technologies AB, Uppsala, Sweden*
ROD FLEMMING • *R&D Department, Promega Corporation, Madison, WI, USA*
BECKY GODAT • *R&D Department, Promega Corporation, Madison, WI, USA*
SAID GOUELI • *R&D Department, Promega Corporation, Madison, WI, USA*
EMILY GRAVENS • *Omega Biologicals, Inc., Bozeman, MT, USA*
MICHAEL J. HALAT • *Omega Biologicals, Inc., Bozeman, MT, USA*
HOVHANNES HAYRAPETYAN • *Marshall B. Ketchum University, Fullerton, CA, USA; Wayne State University, School of Medicine, Detroit, MI, USA*
KEVIN HSIAO • *R&D Department, Promega Corporation, Madison, WI, USA*
JACOB KARP • *Marshall B. Ketchum University, Fullerton, CA, USA*
MARYLA KRAJEWSKA • *Cellestan Immunoquant (CIQ), Inc, Oceanside, CA, USA*
STAN KRAJEWSKI • *Cellestan Immunoquant (CIQ), Inc, Oceanside, CA, USA*
KEVIN KUPCHO • *R&D Department, Promega Corporation, Madison, WI, USA*
DAN LAZAR • *R&D Department, Promega Corporation, Madison, WI, USA*
CHARITHA MADIRAJU • *Marshall B. Ketchum University, Fullerton, CA, USA*
ROBERT S. MATSON • *QuantiScientifics LLC, Orange, CA, USA*
MASARU MATSUDA • *NOF CORPORATION, Life Science Products Division, Tokyo, Japan*
BRYN E. MERRILL • *Chapman University, Orange, CA, USA*
MOTOHIRO MITANI • *NOF CORPORATION, Life Science Products Division, Tokyo, Japan*
OSMANY BLANCO MUÑOZ • *Medicina y Enfermería, Faculta de Ciencias de la Salud, Universidad del Magdalena, Santa Marta, Magdalena, Colombia*
NIDHI NATH • *R&D Department, Promega Corporation, Madison, WI, USA; Bio-Techne, Minneapolis, MN, USA*
TOMOZUMI NODA • *NOF CORPORATION, Life Science Products Division, Tokyo, Japan*
MARTHA O'BRIEN • *R&D Department, Promega Corporation, Madison, WI, USA*
YUSUKE OKAWA • *NOF CORPORATION, Life Science Products Division, Tokyo, Japan*
MARTHA OPPONG • *Marshall B. Ketchum University, Fullerton, CA, USA*
KHOREN SAHAGIAN • *Plasmatreat USA Inc., Hayward, CA, USA*
AMANI SASTRY • *Noorda College of Osteopathic Medicine, Provo, UT, USA*
HIROTAKA SUZUKI • *NOF CORPORATION, Life Science Products Division, Tokyo, Japan*
ANDRZEJ SZCZUDLIK • *Department of Neurology, Jagiellonian University Medical College, Krakow, Poland*
EGLIS TELLEZ-CORRALES • *Marshall B. Ketchum University, Fullerton, CA, USA*
BARBARA TOMIK • *Department of Neurology, Jagiellonian University Medical College, Krakow, Poland*
THAO TRAN • *Marshall B. Ketchum University, Fullerton, CA, USA*
DENONG WANG • *Tumor Glycomics Laboratory, SRI International Biosciences Division, Menlo Park, CA, USA*
HICHAM ZEGZOUTI • *R&D Department, Promega Corporation, Madison, WI, USA*

Chapter 1

Enzyme-Linked Immunosorbent Assay: Types and Applications

Hovhannes Hayrapetyan, Thao Tran, Eglis Tellez-Corrales, and Charitha Madiraju

Abstract

Enzyme-linked immunosorbent assay (ELISA) is an immunological assay widely used in basic science research, clinical application studies, and diagnostics. The ELISA technique relies on the interaction between the antigen (i.e., the target protein) versus the primary antibody against the antigen of interest. The presence of the antigen is confirmed through the enzyme-linked antibody catalysis of the added substrate, the products of which are either qualitatively detected by visual inspection or quantitatively using readouts from either a luminometer or a spectrophotometer. ELISA techniques are broadly classified into direct, indirect, sandwich, and competitive ELISA—all of which vary based on the antigens, antibodies, substrates, and experimental conditions. Direct ELISA relies on the binding of the enzyme-conjugated primary antibodies to the antigen-coated plates. Indirect ELISA introduces enzyme-linked secondary antibodies specific to the primary antibodies bound to the antigen-coated plates. Competitive ELISA involves a competition between the sample antigen and the plate-coated antigen for the primary antibody, followed by the binding of enzyme-linked secondary antibodies. Sandwich ELISA technique includes a sample antigen introduced to the antibody-precoated plate, followed by sequential binding of detection and enzyme-linked secondary antibodies to the recognition sites on the antigen. This review describes ELISA methodology, the types of ELISA, their advantages and disadvantages, and a listing of some multifaceted applications both in clinical and research settings, including screening for drug use, pregnancy testing, diagnosing disease, detecting biomarkers, blood typing, and detecting SARS-CoV-2 that causes coronavirus disease 2019 (COVID-19).

Key words ELISA, Direct ELISA, Indirect ELISA, Sandwich ELISA, Competitive ELISA, COVID-19

1 Introduction

Enzyme-linked immunosorbent assay (ELISA) is an immunological technique extensively used in research and clinical laboratory settings to quantitatively identify a specific protein (i.e., the antigen or biomarker) in a biological matrix while relying on the principle of the specific binding interaction between the antigen and the

Robert S. Matson (ed.), *ELISA: Methods and Protocols*, Methods in Molecular Biology, vol. 2612,
https://doi.org/10.1007/978-1-0716-2903-1_1,

antibody against the antigen of interest [1–12]. ELISA technique commonly involves the following steps: (1) detection of an antigen of interest using a primary antibody that binds to the antigen; (2) a secondary antibody linked to an enzyme that binds to the primary antibody; and (3) a substrate that undergoes hydrolysis—the product of which induces emission of a signal that is detected by (a) change in relative light units (RLUs) using a luminometer, (b) change in the optical density (OD) using a spectrophotometer, and (c) visual observation of the precipitated product, all of which indicate the presence of the protein of interest. Hence, this technique is used in biomedical research or for diagnostic purposes to identify and measure the amount of desired antigen (protein or biomarker) present in the sample or a biological matrix. This chapter on ELISA will provide an overview of the ELISA technique, the historical background on the invention of ELISA, the different types of ELISA, and the application of ELISA in clinical testing and diagnosis.

2 Historical Background of ELISA

The development of today's widely used ELISA, an immunological assay technique, dates back to the 1960s when two research groups simultaneously worked on conceptualizing ELISA and enzyme immunoassay (EIA) techniques via modification of the radioimmunoassay (RIA). The groups were that of Perlman and Engvall at Stockholm University of Sweden and Schuurs and Bauke van Weemen in the Netherlands. The ELISA technique utilizes processes that are characteristic of the human immune response to various antigens and their detection via the use of antibodies. Eva Engvall and Peter Perlman from Sweden first invented the ELISA technique in the 1970s by conjugating their protein of target with the enzyme alkaline phosphatase (AP) [4, 13]. Concurrently, Bauke van Weeman and Anton Schuurs from the Netherlands discovered a similar technique but used the enzyme horseradish peroxidase (HRP) instead [4, 13]. The technical concept of ELISA was developed by modifying the RIA, in which the radioactive iodine-125 used in RIA was replaced with enzymes, AP, and HRP [4]. Although both RIA and ELISA techniques are highly sensitive in detecting the analyte of interest and quantifying the analyte concentration in a sample, RIA involves the use of radioactive molecules, which require special handling that can cause potential health issues [14]. Moreover, the half-life of a radioactive molecule used in RIA is short, thus making RIA a time-sensitive assay [14].

ELISA was first used to measure immunoglobulin G (IgG) in rabbit serum [4]. Shortly after, it was utilized to detect human choriogonadotropin (hCG), a hormone that is produced during pregnancy [4]. Since then, ELISA has been used extensively in various screening tests, including the identification of antigenic components from the Human Immunodeficiency Virus (HIV) and Hepatitis B virus. Since its invention, ELISA has undergone various modifications with the goal to improve the sensitivity of analyte detection, leading to the invention of various types of ELISA, which include direct, indirect, sandwich, and competitive ELISA. Rapid technological advancements saw the development of automated ELISA processing systems to increase reproducibility and efficiency by eliminating the need for manual operations during washing, dispensing, and incubation steps. Despite the emergence of newer scientific methods, ELISA is used extensively in research and laboratory settings even to this day, and this is due to its simplicity and the ability to provide results with high sensitivity, specificity, and selectivity.

3 General ELISA Procedure

ELISA is performed in a 96-well or a 384-well polystyrene plate. The ELISA technique relies on two important elements [1, 2]. First is the specific interaction between the antigen and the antibody. Due to the high fidelity in binding between the antigen and the antibody, any unbound component can be washed away from the assay plates, thus reducing the background noise due to nonspecific binding and improving the accuracy of the results derived from ELISA. However, batch-to-batch variability in antibody production may result in variable results, adding to one of the possible disadvantages of ELISA. Second is the biological property of the assay plates. Polystyrene microtiter plates that form a stable linkage with the bound antigen or antibody without any interference are typically employed for setting up ELISAs.

Types of ELISA

There are different types of ELISA techniques used depending on the specific needs of the researcher or the clinician. These are direct, indirect, sandwich, and competitive ELISA, which vary based on the types of antibodies, antigens, substrates, and experimental conditions used. The selection of a given type of ELISA mainly depends on the analyte of interest, sample source, and purpose of the analysis. The general methodology for any ELISA type described below includes coating, blocking, detection, and a final read. Washing steps are introduced in between to remove any residual unbound components from each step. The four types of ELISA and the underlying principle for each type are listed below. Subsequent sections have the description of different types of ELISA.

1. *Direct ELISA*: The assay plate is coated with the antigen detected by a primary enzyme-conjugated antibody [15, 16].
2. *Indirect ELISA*: The assay plate is coated with the antigen detected by a conjugated primary antibody. An enzyme-labeled antispecies secondary antibody that binds to the primary antibody is then added [17–19].
3. *Sandwich ELISA*: The assay plate is first coated with the primary antibodies; antigens, conjugated primary and enzyme-labeled antispecies secondary antibodies are then consecutively added [19–22].
4. *Competitive ELISA*: This type of ELISA utilizes an uncoated antigen or a mixture of an uncoated antigen plus an unlabeled or an enzyme-conjugated antibody derived from the sample source either of which, respectively, competes for the same immobilized antigen or an unlabeled primary antibody captured on the assay plate [19, 23–25].

3.1 Direct ELISA

Direct ELISA test is performed by first immobilizing the sample containing the antigen or different amounts (or concentrations) of the standard on the surface of each well of a polystyrene 96-well plate [15, 16]. The coated antigen plate is then incubated to allow the formation of a stable linkage between the plate and the antigen. Wash steps are introduced in between each reagent addition step. A blocking buffer is prepared using, for example, one of the following: bovine serum albumin (BSA), serum, and nonfat dry milk. Blocking buffer is then added to cover the remaining unoccupied binding sites or nonspecific binding sites in the 96-well plate. An enzyme-conjugated antibody (HRP-conjugated antibody), specific to the antigen of interest, is added to the wells (Fig. 1A). The commonly used enzymes for labeling the antibodies include AP and HRP, owing to their chromogenic and chemiluminescent properties. The antibody recognizes and adheres to the antigen present in each well. Any unbound antibody is removed during wash steps. Alternatively, a primary detection antibody specific to the coated sample antigen is added (e.g., biotin-labeled target-specific antibody), incubated, and any unbound primary antibody removed during wash steps; secondary antibody covalently linked to an enzyme [e.g., streptavidin-conjugated horseradish peroxidase (S-HRP) secondary antibody that is an antispecies antibody against the primary antibody] is added, incubated, followed by wash steps again to remove any unbound secondary antibody (Fig. 1B).

A substrate [3,3',5,5'-tetramethylbenzidine (TMB) for HRP-based assays and para-nitrophenyl phosphate (pNPP) for AP-based assays] for the antibody-conjugated enzyme is then added that undergoes enzymatic hydrolysis, and the product formed induces a color change, which is detected either as RLU using a luminometer or OD using a spectrophotometer for

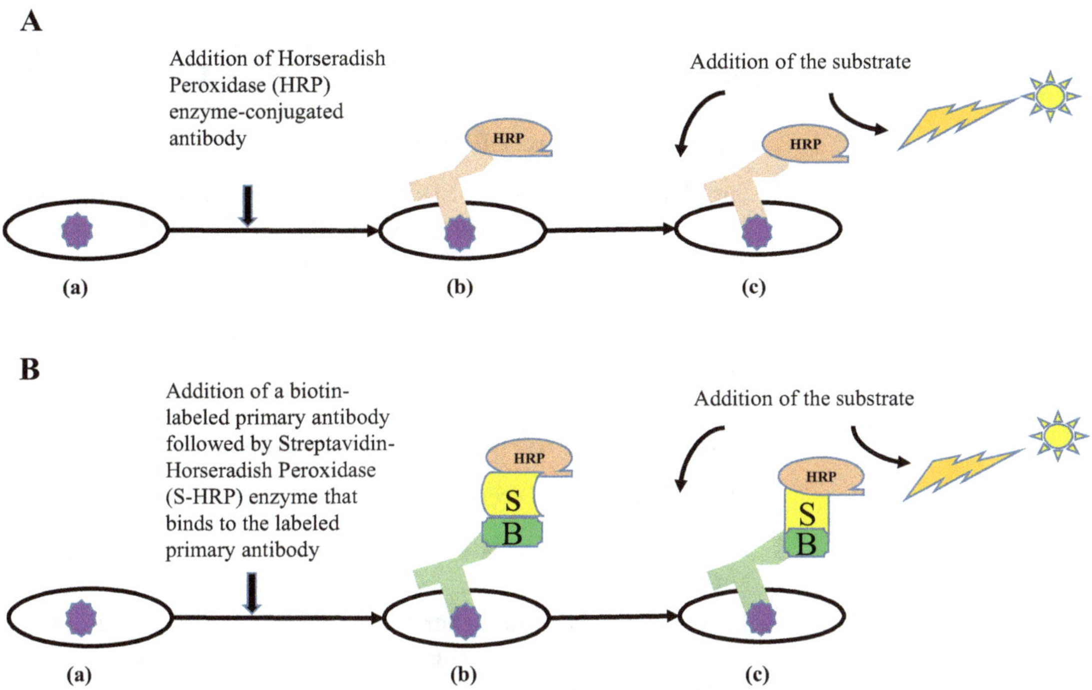

Fig. 1 Schematic illustrations of Direct ELISA. (**a**) Standard or test sample containing the target antigen (a marker of interest) is coated into each well of a 96-well plate. (**b**) A horseradish peroxidase (HRP) enzyme-conjugated antibody (as shown in the top panel of Fig. 1A), or alternatively, biotin-labeled (B) antibody is added that binds to the antigen (purple) that is immobilized into each well of the 96-well plate, followed by the addition of streptavidin-horseradish peroxidase (S-HRP) that binds to the biotin-labeled antibody (as shown in the bottom panel of Fig. 1B). (**c**) Enzymatic hydrolysis of the substrate added into the next step results in a color change that can be detected as either relative light units (RLUs) using a luminometer or optical density (OD) using a spectrophotometer or upon visual observation of the precipitated product if a precipitating blotting solution is used. Data analysis yields qualitative and quantitative information based on the observed change in the intensity of the read-out

colorimetric analysis or even just by visual observation of the precipitated product if a precipitating blotting solution is used. For example, if a chemiluminescent substrate is added, then the substrate undergoes enzymatic hydrolysis (e.g., in the presence of S-HRP), and the resultant product formed is measured using a luminometer. Alternatively, if a TMB blotting solution is added, then this blotting solution, upon S-HRP enzymatic hydrolysis, forms a precipitated product that is assayed by colorimetric (OD) or visual observation by microscopic (followed by ImageJ) analysis. Regardless of the read-out, the values can be analyzed both qualitatively, based on the presence of color, and quantitatively, based on the intensity of the color. Luminescence (based on RLU readings) or absorbance (based on OD readings) values are used to build a standard graph for the known concentrations of the standards from which unknown concentrations of the antigen are determined.

ImageJ analysis helps analyze the precipitated product in case a precipitating blotting solution is used. Graphical analysis is used to determine the antigen concentration present in the sample. Schematic illustrations for direct ELISA are shown in Fig. 1A, B.

Advantages of Direct ELISA

1. Simpler and faster than other ELISA techniques since it uses only one antibody addition step and, hence, fewer steps are involved.
2. Cross-reactivity of secondary antibody is eliminated since this assay type involves primary antibody alone. Using one antibody prevents cross-reaction that is usually seen in the presence of a secondary antibody.
3. Less susceptible to errors because there are fewer steps.

Disadvantages of Direct ELISA

1. Less flexible since each target protein requires a specific primary antibody.
2. Less specific than the other types of ELISA due to the restriction of the assay only to the available primary antibodies against an antigen of interest.
3. Reduced sensitivity if the sample contains multiple proteins and the amount of the protein of interest is not as abundant. Other proteins can adsorb onto the wells and interfere with sensitivity and, hence, can affect the experimental outcome.
4. Minimal signal amplification since only one antibody is used. This can result in low assay sensitivity.

3.2 Indirect ELISA

A sample containing the antigen of interest or a known concentration of the standard is coated or immobilized into each well of a 96-well plate [17, 18]. Wash steps are introduced in between each reagent addition step. A blocking agent (as described in Subheading 3.1) is used to block any uncoated regions within the well and remove any nonspecific binding that may occur during subsequent steps of the assay procedure. A primary antibody specific to the immobilized antigen is added, incubated, and any unbound antibody removed during wash steps. An HRP enzyme-conjugated secondary antibody against the host species of the primary antibody is added (Fig. 2A); or alternatively, a biotin-labeled (B) secondary antibody against the host species of the primary antibody is added, followed by addition of S-HRP that binds to the biotin-labeled antibody (Fig. 2B). Secondary antibody addition steps also include incubation, followed by wash steps again to remove any unbound secondary antibodies. The subsequent step involves the addition of an unprecipitating or a precipitating substrate, which undergoes hydrolysis in the presence of the enzyme-linked secondary antibody, and the resultant product formed is measured in terms of

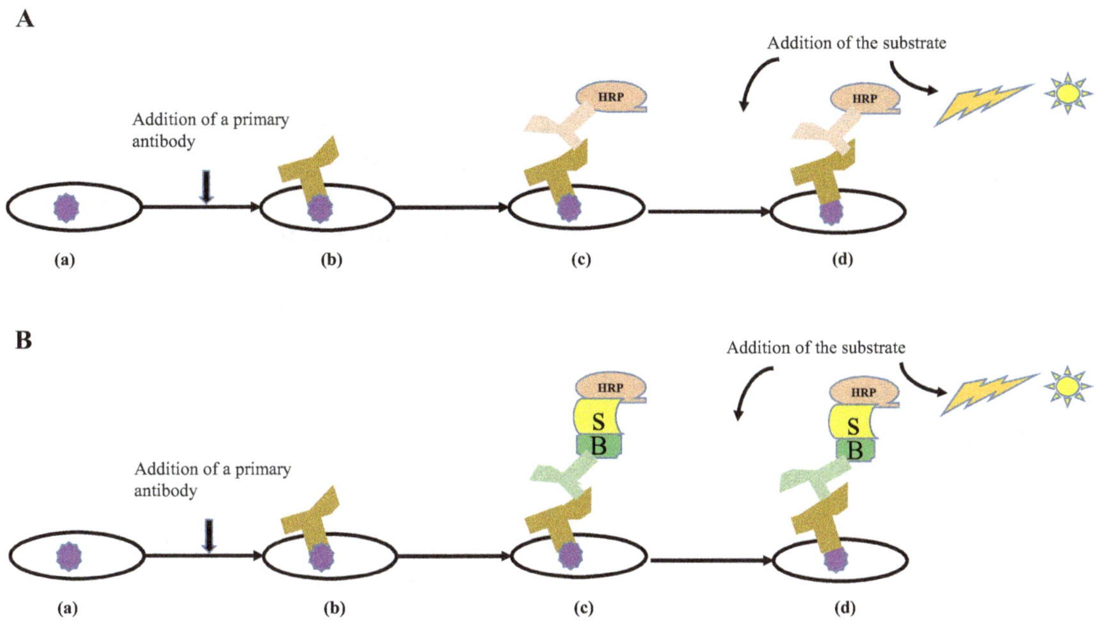

Fig. 2 Schematic illustrations of indirect ELISA. (**a**) Standard or test sample containing the target antigen (a marker of interest) is coated into each well of a 96-well plate. (b) The primary antibody added into each well binds to the immobilized antigen (purple). (**c**) A horseradish peroxidase (HRP) enzyme-conjugated secondary antibody against the host species of the primary antibody is added (as shown in the top panel of Fig. 2A), or alternatively, biotin-labeled (B) secondary antibody against the host species of the primary antibody is added, followed by the addition of streptavidin-horseradish peroxidase (S-HRP) that binds to the biotin-labeled antibody (as shown in the bottom panel of Fig. 2B). (**d**) Enzymatic hydrolysis of the substrate added into the next step results in a color change that can be detected as either relative light units (RLUs) using a luminometer or optical density (OD) using a spectrophotometer or upon visual observation of the precipitated product if a precipitating blotting solution is used. Data analysis yields qualitative and quantitative information based on the observed change in the intensity of the read-out

RLU or OD, or simple image analysis using ImageJ for the precipitated product. As described above for direct ELISA, the RLU, or OD, or ImageJ analysis of an unknown sample can be used to derive the concentration of the unknown samples from the respective standard curve generated upon using known target antigen concentrations. Schematic illustrations of indirect ELISA are shown in Fig. 2A, B.

Advantages of Indirect ELISA

1. Various antigens can be detected using the same species-specific secondary antibody that binds to the species in which the primary antibodies are raised.
2. Signal amplification is achievable since a secondary antibody is used. This can result in higher assay sensitivity than direct ELISA.

Disadvantages of Indirect ELISA

1. Like in direct ELISA, the method of antigen immobilization is not specific. When a biological matrix (serum, plasma, CSF, etc.) is used as the sample source for the test antigen, some nonspecific proteins in the sample may adhere to the wells of a microtiter plate. Additionally, if the target antigen levels are low and/or if nonspecific proteins are in abundance, indirect ELISA may not work.
2. Cross-reactivity of secondary antibodies with nonspecific antigens is possible since this assay type involves not only primary antibodies but also secondary antibodies.

3.3 Sandwich ELISA

The target-specific capture antibody is precoated into each well of a 96-well plate, and the precoated plate is used for sandwich-type ELISA [20–22]. Prior to testing the standards or test samples, precoated wells are rinsed with wash buffer and blocked with a blocking agent (as mentioned in Subheading 3.1) to prevent any nonspecific binding of the standards or test samples added in the next step. Wash steps are introduced in between each reagent addition step. Known concentrations of standard or test samples (that are the antigens) added to each well are incubated for a given amount of time (according to the protocol) to enable target antigens present in the standards or test samples to bind to the capture antibody. Here, the antigen needs to have at least two antigenic sites for the formation of the "sandwich" in the subsequent step. The unbound antigens (standards or test samples) from each well are washed out during wash steps. An enzyme-conjugated antibody (as shown in Fig. 3A) is added to complete the "sandwich" in case of a type of direct "sandwich" ELISA, and this step is followed by wash steps and then addition of the substrate that undergoes enzymatic hydrolysis as described in subsequent steps below (and as shown in the top panel of Fig. 3A).

Alternatively, in yet another type of direct "sandwich" ELISA technique, a labeled detection (biotinylated) antibody is added (as shown in Fig. 3B) or in case of an indirect "sandwich" ELISA method, an unlabeled primary antibody is added (as shown in Fig. 3C), either of them bind to the captured antigen to complete the "sandwich." A description of direct versus indirect "sandwich" ELISA techniques is provided below. Regardless of whether it is a direct or an indirect "sandwich" ELISA method, the unbound labeled detection or unlabeled primary antibody is washed off during the wash steps. This is followed by the addition of S-HRP that binds to the biotin-labeled (B) antibody (Fig. 3B); alternatively, an HRP-conjugated secondary antibody against the host species of the primary antibody is added that binds to the primary antibody (Fig. 3C). The unbound S-HRP or enzyme-linked antibody conjugate is washed out during wash steps.

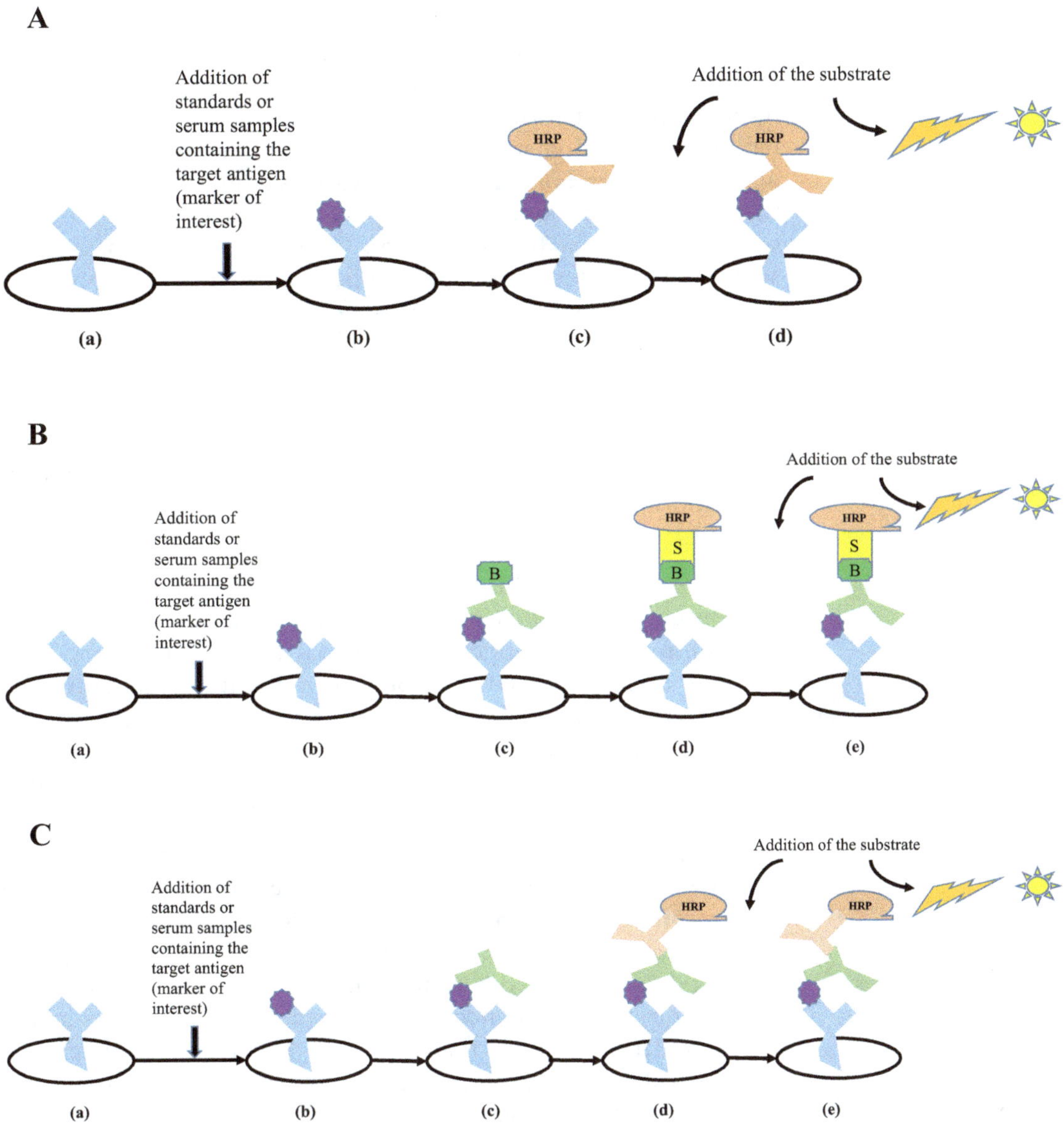

Fig. 3 Schematic illustrations of sandwich ELISA. (**a**) Capture antibody bound or precoated into each well of a 96-well plate. (**b**) Target antigen (purple) binds to the capture antibody precoated into each well of the 96-well plate. (**c**) A horseradish peroxidase (HRP) enzyme-conjugated antibody is added that binds to the target antigen bound to the capture antibody, thus forming a "sandwich" followed by addition of the substrate that undergoes enzymatic hydrolysis as described in subsequent steps below (as shown in the top panel of Fig. 3A), or alternatively, a biotinylated (B) detection antibody is added that binds to the target antigen bound to the capture antibody, thus forming a "sandwich" (as shown in the middle panel of Fig. 3B); or a primary antibody is added that binds to the target antigen bound to the capture antibody, thus forming a "sandwich" (as shown in the bottom panel of Fig. 3C). (**d**) Streptavidin-horseradish peroxidase (S-HRP) is then added that binds to the biotin-labeled (as shown in the middle panel of Fig. 3B), or alternatively, a HRP-conjugated secondary antibody against the host species of the primary antibody is added that binds to the primary antibody (as shown in the bottom panel of Fig. 3C). (**e**) Enzymatic hydrolysis of the substrate added into the next step results in a color change that can be detected as either relative light units (RLUs) using a luminometer or optical density (OD) using a spectrophotometer or upon visual observation of the precipitated product if a precipitating blotting solution is used. Data analysis yields qualitative and quantitative information based on the observed change in the intensity of the read-out

A chemiluminescent substrate is then added that undergoes enzymatic hydrolysis (e.g., in the presence of S-HRP or HRP), and the resultant product formed is measured in terms of RLU using a luminometer. Alternatively, a blotting solution is added, which upon S-HRP or HRP enzymatic hydrolysis precipitates out, and the precipitated product is assayed by colorimetric (OD) or microscopic (followed by ImageJ) analysis. The RLU, or OD, or ImageJ analysis of an unknown sample can be compared to the respective standard curve generated upon using known target antigen concentrations. Schematic illustrations of sandwich ELISA are shown in Fig. 3A–C.

Depending on the type of detection antibodies, sandwich ELISA can be classified as (1) direct sandwich ELISA and (2) indirect sandwich ELISA. A brief description of the differences in detection methods for direct sandwich ELISA and indirect sandwich ELISA is provided below:

1. Direct sandwich ELISA involves either the addition of the same capture antibody (used for precoating the plate prior to the addition of the antigen), except that the capture antibody is now enzyme-labeled or biotinylated for it to function as a detection antibody or a different enzyme- or biotin-labeled detection antibody (either from the same species or different species) that still binds specifically to the bound antigen (Fig. 3A, B).
2. Indirect sandwich ELISA involves three antibodies, one capture antibody that is used to precoat the 96-well plate and that captures the added antigen, one detection antibody that binds to the other antigenic site on the antigen, thus completing the "sandwich," and the third antibody that is an antispecies enzyme-labeled secondary antibody (e.g., HRP enzyme-linked secondary antibody) that specifically binds to the detection antibody. The enzyme-labeled secondary antibody needs to be from another species to prevent it from binding to the capture antibody. However, a single enzyme-labeled conjugate antibody can be used to identify the binding of the detection antibodies used against a variety of antigens, which makes the indirect sandwich ELISA advantageous compared to direct sandwich ELISA (Fig. 3C).

Advantages of Sandwich ELISA

1. Unlike direct or indirect ELISA, the method of antibody immobilization is specific to the sample antigen of interest. When a biological matrix (serum, plasma, CSF, etc.) is used as the sample source for the test antigen, the sample antigen specifically binds to the capture antibody adhered to the wells of the 96-well microtiter plate.

2. Coating the plate with a purified capture antibody specific to the sample antigen eliminates the necessity to purify the antigen from a pool of other antigens (present in a biological matrix), thus making this type of ELISA simple, sensitive, and specific.
3. Sandwich ELISA is especially useful when antigens are in low concentration and not in abundance or when antigens are contaminated with other nonspecific antigens or proteins.
4. Like indirect ELISA, signal amplification is achievable since a secondary antibody is used. This can result in higher assay sensitivity than direct ELISA.

Disadvantages of Sandwich ELISA

1. Cross-reactivity of secondary antibodies is possible since this assay type involves more than one antibody: capture and detection antibodies and, if needed, an enzyme-conjugated secondary antibody.
2. The antigen needs to have two antigen-binding sites, one for the capture antibody and the other for the detection antibody.
3. One limitation of direct sandwich ELISA is that this method can be of use only where single-species antiserum is available and when antigen does not bind to the wells of the microtiter plate.
4. One limitation of indirect sandwich ELISA is that this method also requires the production of enzyme-labeled conjugate antibodies from another species to prevent them from binding to the capture antibody.
5. It is simpler but slower than direct ELISA technique since it uses two or more antibody addition steps. It is more susceptible to error because there are more steps.

3.4 Competitive ELISA

Competitive ELISA is characterized by competition between the precoated or immobilized antigen present in each well of a 96-well plate and the sample antigen-primary antibody mix added into each well of the same 96-well plate [23–25]. Competitive ELISA is configured depending upon whether the analyte is an antigen or an antibody. Various competitive ELISA subtypes exist that are modified versions of direct ELISA, indirect ELISA, and sandwich ELISA described in Subheadings 3.1, 3.2, and 3.3. Accordingly, subtypes of competitive ELISA include the following based on whether the antigen or the antibody component in the original ELISA method is set up for competition: (1) direct ELISA–antigen competition, (2) direct ELISA–antibody competition, (3) indirect ELISA–antigen competition, (4) indirect ELISA–antibody competition, (5) direct sandwich ELISA–antigen or antibody competition, and (6) indirect sandwich ELISA–antigen or antibody competition. A detailed description of each competitive ELISA subtype is beyond the scope of this chapter. However, the following section has a brief description of the underlying principle for each of the broadly classified subtypes of competitive ELISA, which include (1) direct ELISA–antigen competition, (2) direct ELISA–antibody competition, (3) indirect ELISA–antigen competition, and (4) indirect ELISA–antibody competition.

1. Direct ELISA–antigen competition: Competition for the antigen is set up upon addition of a pretitrated mixture of sample antigen (with the same antigenic sites as that of the immobilized antigen) and enzyme-conjugated or labeled (biotinylated) antibody. The added sample antigen in liquid phase competes with the immobilized antigen for binding to the enzyme-conjugated or labeled (biotinylated) antibody against the immobilized antigen (Fig. 4A).
2. Direct ELISA–antibody competition: Competition for the pretitrated enzyme-conjugated antibody binding to the immobilized antigen is set up upon addition of the serum/plasma sample containing the antibodies against the immobilized antigen. The added sample antibodies compete with the pretitrated enzyme-conjugated detection antibodies for binding to the immobilized antigen.
3. Indirect ELISA–antigen competition: Competition for the antigen is set up upon addition of a pretitrated mixture of sample antigen (that has the same antigenic-binding sites as the immobilized antigen) and the primary antibody against the immobilized antigen. The added sample antigen competes with the immobilized antigen and detected upon addition of an enzyme-conjugated (Fig. 4B) or a labeled (biotinylated) (Fig. 4C) antispecies antibody.
4. Indirect ELISA–antibody competition: Competition for the pretitrated primary antibody binding to the immobilized antigen is set up upon addition of the serum/plasma sample containing the antibodies (from another species as the primary antibody) against the immobilized antigen. The added sample antibodies compete with the primary antibody for binding to the immobilized antigen and detected upon addition of antispecies labeled or an enzyme-conjugated detection antibody.

The following description is applicable to the two types involving antigen competition, namely, direct ELISA–antigen competition and indirect ELISA–antigen competition. However, the same format modified for antibody-based competition ELISAs as well.

ELISAs configured for competition against an antigen involve addition to the precoated immobilized antigen (present in each well of a 96-well plate) of a premix of the sample antigen and HRP-conjugated or biotin-labeled antibody (if it is direct ELISA–

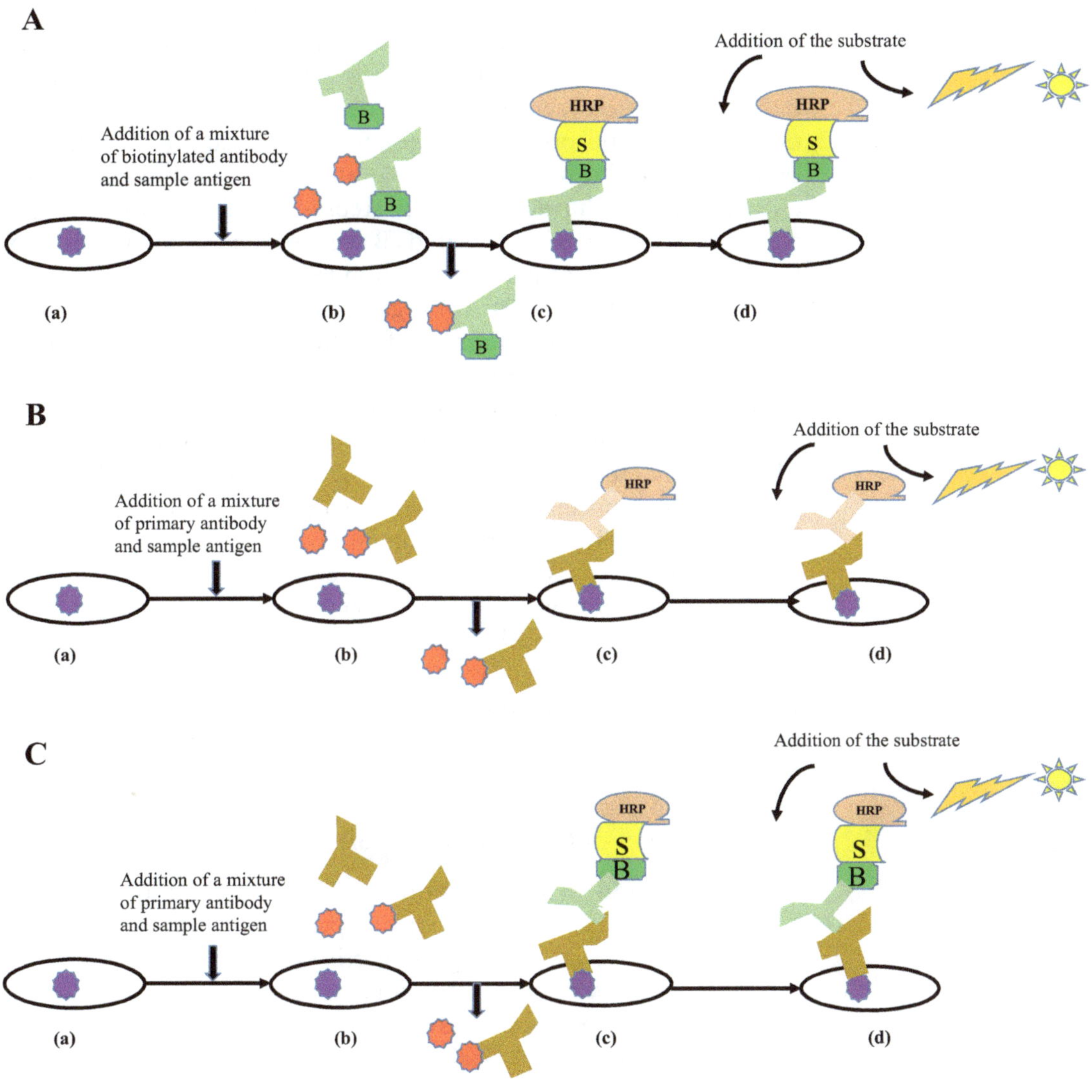

Fig. 4 Schematic illustrations of competitive ELISA. (**a**) Control antigen is immobilized into each well of a 96-well plate. (**b**) The immobilized antigen (purple) competes with the sample antigen (red) for binding to the primary biotinylated (B) antibody (top panel of Fig. 4A), or alternatively, the immobilized antigen (purple) competes with the sample antigen (red) for binding to the primary antibody (middle panel of Fig. 4B and bottom panel of Fig. 4C). (**c**) Streptavidin-horseradish peroxidase (S-HRP) is then added that binds to the biotinylated (B) detection antibody (top panel of Fig. 4A), or alternatively, an HRP enzyme-conjugated secondary antibody against the host species of the primary antibody is added (as shown in the middle panel of Fig. 4B), or biotin-labeled (B) secondary antibody against the host species of the primary antibody is added, followed by the addition of S-HRP that binds to the biotin-labeled antibody (as shown in the bottom panel of Fig. 4C). (**d**) Enzymatic hydrolysis of the substrate added into the next step results in a color change that can be detected as either relative light units (RLUs) using a luminometer or optical density (OD) using a spectrophotometer or upon visual observation of the precipitated product if a precipitating blotting solution is used. Data analysis yields qualitative and quantitative information based on the observed change in the intensity of the read-out

antigen competition as shown in Fig. 4A) or a premix of sample antigen and unlabeled primary antibody ((if it is indirect ELISA–antigen competition as shown in Fig. 4B). Wash steps are introduced in between each reagent addition steps. There is a competition between the precoated antigen and the sample antigen for binding to antibody. The unbound antibody and sample antigen are washed off during the wash steps. This is followed by the addition of S-HRP if biotinylated (B) antibody is used in the premix for direct ELISA–antigen competition (Fig. 4A).

Alternatively, an HRP-conjugated secondary antispecies antibody that binds to the primary detection antibody (as shown in Fig. 4B) or a biotinylated antispecies detection antibody followed by S-HRP (as shown in Fig. 4C) is added. The secondary antibody is added and incubated for it to bind to the primary antibody, and the unbound secondary antibody is washed off during the wash steps. A chemiluminescent substrate is then added that undergoes enzymatic hydrolysis and the resultant product formed is measured in terms of RLU using a luminometer. Alternatively, a blotting solution is added, which upon enzymatic hydrolysis precipitates out, and the precipitated product is assayed by colorimetric (OD) or microscopic (followed by ImageJ) analysis. The RLU, or OD, or ImageJ analysis of an unknown sample can be compared to the respective standard curve generated upon using known target antigen concentrations. Schematic illustrations of competitive ELISA are shown in Fig. 4A–C.

Advantages of Competitive ELISA

1. The method of competition between sample antigen versus immobilized antigen-antibody is specific to the sample antigen of interest. When a biological matrix (serum, plasma, or CSF) is used as the sample source for the test antigen, the sample antigen can specifically compete with the antigen-capture antibody adhered to the wells of the 96-microtiter plate.
2. Other advantages listed under Subheadings 3.1, 3.2, and 3.3 hold for competitive ELISAs depending on the subtype of competitive ELISA.

Disadvantages of Competitive ELISA

1. Competition between sample antigen and immobilized antigen may not be specific. When a biological matrix (serum, plasma, or CSF) is used as the sample source for the test antigen, some contaminated or nonspecific proteins (that are abundant) in the sample matrix may compete with the coated antigen or antibody present in the wells of a microtiter plate.
2. Other disadvantages listed under Subheadings 3.1, 3.2, and 3.3 hold for competitive ELISAs depending on the subtype of competitive ELISA.

4 Applications of ELISA and Clinical Significance

There are multifaceted applications to the ELISA technique. ELISA has been used in very different settings and has a wide range of applications. ELISA is extensively used as a screening tool for the detection of infectious diseases, drug use, quantifying specific substances to aid in the diagnosis, monitoring, and assessment of pharmacodynamics and tumor markers. Additionally, it is utilized in pregnancy testing, blood typing, and in a variety of diagnostic and research settings (*see* Table 1 for some of these applications). For example, owing to their sensitivity, ELISAs are extensively used in research laboratories as screening tools for infections caused by pathogenic microbes [e.g., *Haemophilus influenzae*, Influenza virus strains, HIV, Hepatitis B, etc.], measuring hormone levels (i.e., hCG), detecting potential food allergens, detecting and measuring tumor markers in cancers, such as prostate-specific antigen and carcinoembryonic antigen, and screening for the presence of antibodies in the blood [9, 13, 26, 27].

Currently, one of the most significant applications of ELISA is its application in COVID-19 serology testing. Serology or antibody testing for COVID-19 determines whether an individual has had a past COVID-19 infection by detecting the presence of SARS-CoV-2 antibody developed from an adaptive immune response.

Table 1
Applications of ELISA

Applications	Examples
Detect and measure the presence of antibodies in the blood [30–33]	Autoantibodies and antibodies against infectious disease (antibacterial, antiviral, antifungal)
Detect and estimate the levels of tumor markers [34–36]	Prostate-specific antigen (PSA) Carcinoembryonic antigen (CEA)
Detect and estimate hormone levels [37, 38]	Luteinizing hormone Follicular-stimulating hormone Prolactin Testosterone Human chorionic gonadotropin (hCG)
Screening for possible viral contaminants in blood [39, 40]	Anti-HIV-1/2 Anti-HCV HBsAg
Detecting drug abuse [41, 42]	Amphetamine, 3,4-methylenedioxymethamphetamine Cocaine Cannabinoids Benzoylecgonine
Allergy diagnosis and research [12, 43]	

Serological testing is critical for determining the actual number of COVID-19 cases since it confirms the positive cases even in those who are asymptomatic and are unable to get the molecular tests. ELISA serological tests are performed using a 96-well polystyrene plate coated with purified SARS-CoV-2 recombinant proteins, to which the serum samples and the antihuman IgG antibodies coated with HRP are subsequently added [28]. The detected level of IgG antibodies against SARS-CoV-2 correlates with the level of neutralizing antibodies present in an individual [28]. In one of the first research studies that were conducted at the Mount Sinai Hospital in New York City in 2020, transfusion of convalescent plasma containing SARS-CoV-2 neutralizing antibodies showed an increase in survival rate and a decrease in oxygen therapy dependence among the recipients [29]. Although a larger randomized trial is needed to further consolidate the data, COVID-19 neutralizing antibodies showed promising results, and this approach is one of the treatment options for COVID-19 [29]. As the world continues to battle with the pandemic and as scientific techniques evolve, ELISA will continue to make its impact on laboratory research and clinical settings.

References

1. Butler JE (2000) Enzyme-linked immunosorbent assay. J Immunoass 21(2–3):165–209
2. Crowther JR (1995) ELISA. Theory and practice. Methods Mol Biol (Clifton, N.J.) 42:1–218
3. Engvall E (2010) The ELISA, enzyme-linked immunosorbent assay. Clin Chem 56(2): 319–320
4. Engvall E, Perlmann P (1971) Enzyme-linked immunosorbent assay (ELISA). Quantitative assay of immunoglobulin G. Immunochemistry 8(9):871–874
5. Hornbeck P (1992) Enzyme-Linked Immunosorbent Assays. Curr Protoc Immunol 1: 2.1.1–2.1.22
6. Hornbeck P, Winston SE, Fuller SA (1991) Enzyme-linked immunosorbent assays (ELISA). Curr Protoc Mol Biol 15: 11.2.1–11.2.22
7. Hornbeck PV (2015) Enzyme-linked immunosorbent assays. Curr Protoc Immunol 110: 2.1.1–2.1.23
8. Kohl TO, Ascoli CA (2017) Immunoassays. Cold Spring Harbor Protoc 7:pdb.top093690
9. Lequin RM (2005) Enzyme immunoassay (EIA)/enzyme-linked immunosorbent assay (ELISA). Clin Chem 51(12):2415–2418
10. Reen DJ (1994) Enzyme-linked immunosorbent assay (ELISA). Methods Mol Biol (Clifton, N.J.) 32:461–466
11. Aydin S (2015) A short history, principles, and types of ELISA, and our laboratory experience with peptide/protein analyses using ELISA. Peptides 72:4–15
12. Konstantinou GN (2017) Enzyme-linked immunosorbent assay (ELISA). Methods Mol Biol (Clifton, N.J.) 1592:79–94
13. Alhajj M, Farhana A (2022) Enzyme linked immunosorbent assay. In StatPearls, Treasure Island (FL)
14. Grange RD, Thompson JP, Lambert DG (2014) Radioimmunoassay, enzyme and non-enzyme-based immunoassays. Br J Anaesth 112(2):213–216
15. Kohl TO, Ascoli CA (2017) Direct and indirect cell-based enzyme-linked immunosorbent assay. Cold Spring Harbor Protoc 5:pdb.prot093732
16. Lin AV (2015) Direct ELISA. Methods Mol Biol (Clifton, N.J.) 1318:61–67
17. Kohl TO, Ascoli CA (2017) Indirect immunometric ELISA. Cold Spring Harbor Protoc 5: pdb.prot093708
18. Lin AV (2015) Indirect ELISA. Methods Mol Biol (Clifton, N.J.) 1318:51–59
19. Gan SD, Patel KR (2013) Enzyme immunoassay and enzyme-linked immunosorbent assay. J Invest Dermatol 133(9):e12
20. Kohl TO, Ascoli CA (2017) Immunometric antibody Sandwich enzyme-linked

immunosorbent assay. Cold Spring Harbor Protoc 6:pdb.prot093716

21. Kohl TO, Ascoli CA (2017) Immunometric double-antibody Sandwich enzyme-linked immunosorbent assay. Cold Spring Harbor Protoc 6:pdb.prot093724
22. Stanker LH, Hnasko RM (2015) A double-Sandwich ELISA for identification of monoclonal antibodies suitable for Sandwich immunoassays. Methods Mol Biol (Clifton, N.J.) 1318:69–78
23. Kohl TO, Ascoli CA (2017) Direct competitive enzyme-linked immunosorbent assay (ELISA). Cold Spring Harbor Protoc 7:pdb.prot093740
24. Kohl TO, Ascoli CA (2017) Indirect competitive enzyme-linked immunosorbent assay (ELISA). Cold Spring Harbor Protoc 7:pdb.prot093757
25. Makarananda K, Weir LR, Neal GE (1998) Competitive ELISA. Methods Mol Biol (Clifton, N.J.) 80:155–160
26. Plested JS, Coull PA, Gidney MAJ (2003) ELISA. In: Herbert MA, Hood DW, Moxon ER (eds) Haemophilus influenzae protocols. Humana Press, Totowa, pp 243–261
27. Zhang A, Jin M, Liu F et al (2006) Development and evaluation of a DAS-ELISA for rapid detection of avian influenza viruses. Avian Dis 50(3):325–330
28. Mendoza R, Silver M, Zuretti AR et al (2021) Correlation of automated chemiluminescent method with enzyme-linked immunosorbent assay (ELISA) antibody titers in convalescent COVID-19 plasma samples: development of rapid, cost-effective semi-quantitative diagnostic methods. J Blood Med 12:157–164
29. Liu STH, Lin HM, Baine I et al (2020) Convalescent plasma treatment of severe COVID-19: a propensity score-matched control study. Nat Med 26(11):1708–1713
30. Alsaed OS, Alamlih LI, Al-Radideh O et al (2021) Clinical utility of ANA-ELISA vs ANA-immunofluorescence in connective tissue diseases. Sci Rep 11(1):8229
31. Dighiero G, Magnac C, de Saint Martin J et al (1987) Detection of anti-mitochondrial antibodies by ELISA and Western-blot techniques and identification by one and two-dimensional gel electrophoresis of M2 target antigens. Clin Exp Immunol 70(3):640–648
32. Liu H, Norman GL, Shums Z et al (2010) PBC screen: an IgG/IgA dual isotype ELISA detecting multiple mitochondrial and nuclear autoantibodies specific for primary biliary cirrhosis. J Autoimmun 35(4):436–442
33. Roth KDR, Wenzel EV, Ruschig M et al (2021) Developing recombinant antibodies by phage display against infectious diseases and toxins for diagnostics and therapy. Front Cell Infect Microbiol 11:697876
34. El-Masry S, El-Sayed IH, Lotfy M et al (2007) Utility of slot-blot-ELISA as a new, fast, and sensitive immunoassay for detection of carcinoembryonic antigen in the urine samples of patients with various gastrointestinal malignancies. J Immunoassay Immunochem 28(2): 91–105
35. Panini NV, Messina GA, Salinas E et al (2008) Integrated microfluidic systems with an immunosensor modified with carbon nanotubes for detection of prostate specific antigen (PSA) in human serum samples. Biosens Bioelectron 23(7):1145–1151
36. Yin Y, Cao Y, Xu Y et al (2010) Colorimetric immunoassay for detection of tumor markers. Int J Mol Sci 11(12):5077–5094
37. Al-Dujaili EA (2006) Development and validation of a simple and direct ELISA method for the determination of conjugated (glucuronide) and non-conjugated testosterone excretion in urine. Clin Chim Acta 364(1–2):172–179
38. Kuo HT, Yeh JZ, Wu PH et al (2012) Application of immunomagnetic particles to enzyme-linked immunosorbent assay (ELISA) for improvement of detection sensitivity of HCG. J Immunoassay Immunochem 33(4):377–387
39. Beelaert G, Fransen K (2010) Evaluation of a rapid and simple fourth-generation HIV screening assay for qualitative detection of HIV p24 antigen and/or antibodies to HIV-1 and HIV-2. J Virol Methods 168(1–2): 218–222
40. Workowski KA, Bolan GA (2015) Centers for disease, C.; prevention, sexually transmitted diseases treatment guidelines, 2015. MMWR Recomm Rep 64(RR-03):1–137
41. Agius R, Nadulski T (2014) Utility of ELISA screening for the monitoring of abstinence from illegal and legal drugs in hair and urine. Drug Test Anal 6(Suppl 1):101–109
42. Olivieri B, Maric M, Bridge C (2018) Determining the effects of adulterants on drug detection via enzyme-linked immunosorbent assay and adulterant tests strips. Drug Test Anal 10(9):1383–1393
43. Weng X, Gaur G, Neethirajan S (2016) Rapid detection of food allergens by microfluidics ELISA-Based Optical sensor. Biosensors (Basel) 6(2):24

Chapter 2

ELISA Essentials: Surfaces, Antibodies, Enzymes, and Substrates

Robert S. Matson

Abstract

The enzyme-linked immunosorbent assay is a powerful analytical tool for the assessment of the kind and quantity of specific analytes found within a biological sample. It is based upon the exceptional specificity of antibody recognition of its cognate antigen and the power of enzyme-mediated signal amplification for sensitivity. However, development of the assay is not without challenges. Here, we describe the essential components and features necessary to successfully prepare and carry out the ELISA format.

Key words Antibody, Antigen, Enzyme, Immunosorbent, Microplates, Loading, Interference, HAAA, Streptavidin–biotin, HRP, ALP, Inhibitors, Hook effect, Coating, Blocking, Diluent, Matrix effect

1 Introduction

While the application of the enzyme-linked immunosorbent assay (ELISA) method is widespread, the choice of enzymes and their respective substrates has been somewhat limited in scope. What factors are taken into consideration when selecting enzymes suitable for use in an immunoassay? What enzyme substrates are available as signaling reagents for colorimetric, fluorescence, chemiluminescent detection? To understand the requirements necessary for utilization of the enzyme for ELISA, we first need to examine the processes involved in carrying out the immunoassay.

The essential elements of the immunoassay (Fig. 1) are a solid phase (the immunosorbent), an antigen (analyte in sample) associated directly or indirectly to the solid phase, an antibody that recognizes (binds to) the sequestered antigen, and an enzyme that confirms the binding event (by catalysis of the enzyme substrate to a detectable product). However, there are many nuances to contend with.

Robert S. Matson (ed.), *ELISA: Methods and Protocols*, Methods in Molecular Biology, vol. 2612,
https://doi.org/10.1007/978-1-0716-2903-1_2,

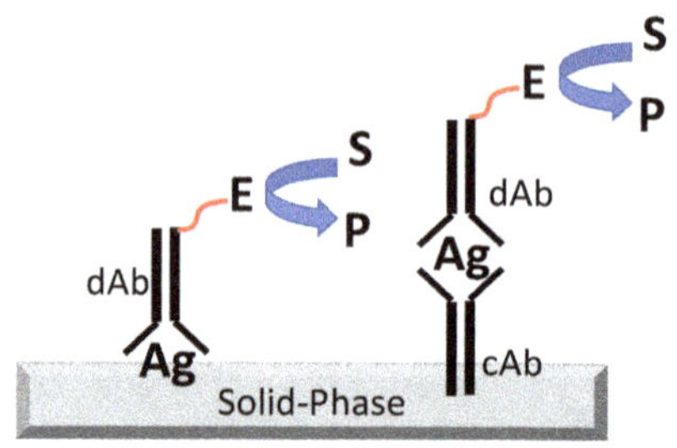

Fig. 1 Essential elements of ELISA. Antigen (Ag) or analyte is directly adsorbed to the solid phase or conversely a surface-bound capture antibody (cAb) is used to sequester (Ag). Next, a detection antibody (dAb) binds to (Ag). An enzyme (E) tethered to the dAb catalyzes the conversion of a substrate (S) to a signal product (P), thereby confirming the presence or absence of the analyte (Ag)

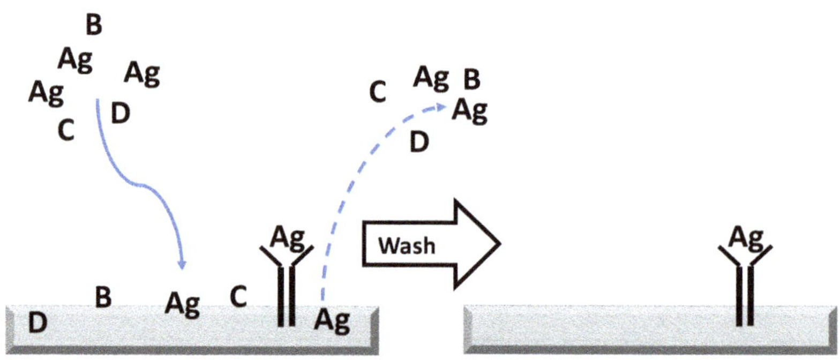

Fig. 2 Removal of loosely bound antigen. The sample not only contains the antigen (analyte) but also a host of other biomolecules (B, C, D, etc.). These nonspecific biomolecules along with loosely bound antigen need to be removed from the surface prior to performing ELISA

First, the sample containing the analyte (antigen) contains other biomolecules that can adsorb to the solid phase. Moreover, not all the antigens will necessarily bind to the same extent. For these reasons, the surface usually requires rinsing to remove loosely bound antigen and other adsorbed biomolecules (Fig. 2). In most cases, it is also necessary to coat the surface (block) with a blocking agent such as bovine serum albumin (BSA) that will substantially reduce nonspecific binding (Fig. 3). Otherwise, interferences with the assay can occur. For example, loosely bound antigen could release back into solution and bind to the added detection antibody, thereby reducing the signal. Other biomolecules that adsorb to the surface may become a site for nonspecific binding of the detection antibody, enzyme, or signaling reagents, leading to a false-positive signal. There are also matrix effects. For example, it is known that carbohydrates, immunoglobulins, or other proteins may bind to the capture antibody or detection antibody, thereby reducing the binding of the antigen; or the presence of small

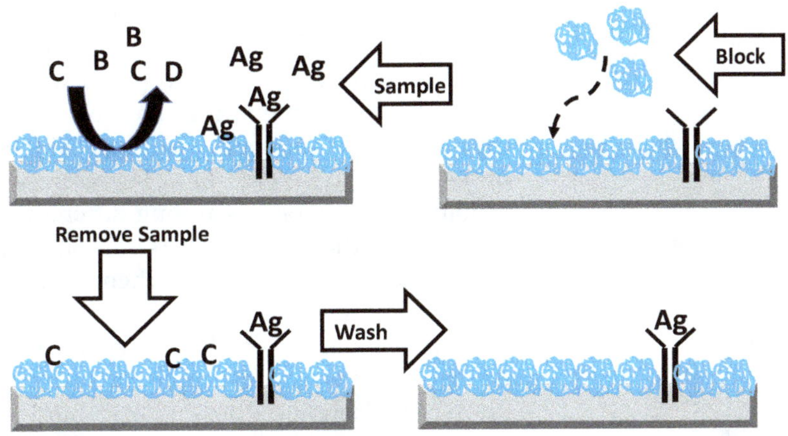

Fig. 3 Addition of a blocking agent. After coating the capture antibody, the surface is treated with a blocking agent that further reduces the nonspecific binding. In this case, the blocker reduced the binding of B, D, and Ag while C was removed during the wash step

molecules, metals, or pH shifts may affect the performance of the enzyme. There can be differences in these effects among different matrices such as urine, serum, or plasma.

In summary, the ELISA physical platform (e.g., microtiter plate) involves selection of a suitable solid phase, immobilization of the capture antibody, and subsequent blocking of the surface to reduce nonspecific binding. The development of immunoassay itself further requires careful selection and validation of the capture and detection antibodies together with determination of the most suitable detection mode that will achieve the highest sensitivity and specificity. Here, the choice of the enzyme conjugate and substrates is important. To be successful in validating ELISA, it is necessary to overcome nonspecific binding, interferences, and matrix effects.

2 Immunosorbent

Selection of the solid phase or immunosorbent is most critical to the success of the development of an immunoassay. Most ELISAs are conducted using plastic microtiter plates. However, not all microplates are suitable for this purpose. Those used for cell culturing should generally be avoided since they most often lead to high nonspecific backgrounds. Likewise, glass-bottomed plates used for spectral analysis are unsuitable since surface binding is relatively low for immobilization of proteins. There are also "nonbinding" microplates that should not be used. However, these may be nonbinding of DNA but may bind protein. If in doubt, check the adsorption of antibody to the surface.

Microplates specifically designed to be used for ELISA are readily available. These are offered in various plastic materials (commonly polystyrene) that have been surface modified into various grades based upon binding capacity for IgG adsorption: low binding, 200–250; moderate binding, 350–500; and high binding, 600–650 ng IgG/cm^2 [1]. There are differences in the passive binding capacity ranges among different manufacturers. If, on the other hand, the assay involves immobilization of other proteins to the surface (antigen-down), then selection of the plate (surface) may be more dependent upon the relative hydrophobicity or hydrophilic nature of the materials. For example, hydrophobic proteins bind well to hydrophobic surfaces while those with hydrophobic/hydrophilic domains may bind better to a more hydrophilic surface. Finally, it should be noted that plates having covalent linking chemistries or adsorbed coatings such as streptavidin (SA) are also available in plastics such as polystyrene, polypropylene, and, most recently, cyclic olefin copolymer (COC). In most circumstances, plates designed for passive adsorption are sufficient and more cost-effective.

3 Antibodies

Not all antibodies are created equally. One should not assume that the same antibody obtained from one vendor will behave the same as that sourced from a different vendor. There can be significant differences in quality, buffer composition, carrier, protein concentration, subclass, epitope recognition, clonality, host, physical state, antigen binding, and thermal stability. Compare vendor data sheets or certificates of analysis (CoA). Get to know your antibody.

It is advisable to source from at least 2–3 different suppliers for evaluation. If a sandwich ELISA is required, then also source from multiple vendors. Keep in mind that the capture antibody (surface bound) and detection antibody (labeled) should recognize different epitopes of the antigen. In case polyclonal antibodies are selected, it is important to determine whether there is any cross-species binding between capture and detection antibodies. This is a common problem and can occur due to different circumstances for both monoclonal and polyclonal pairs (Fig. 4). The presence of heterophile antibodies is often associated with human antianimal antibodies (HAAA) since about 40% of the general population [2] produce antibodies to HAAA, e.g., antimouse or human antimouse antibody (HAMA). The issue of cross-reactivity is further compounded when utilizing microarray-based sandwich assays [3].

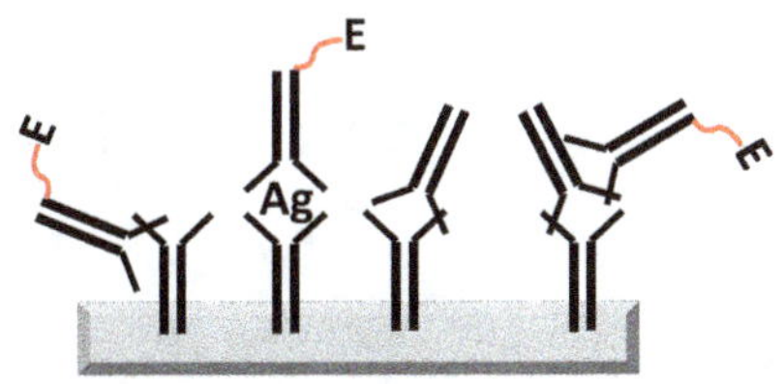

Fig. 4 Cross-species reactivity. Sources of heterophile antibody (or cross-reacting antibody) can come from the detection antibody recognition of an epitope of the capture antibody (**a**) leading to a false-positive signal, or (**b**) where an endogenous antibody from the sample binds to the capture antibody, leading to a false-negative signal. Also possible is the bridging of the detection antibody, which recognizes a cross-reacting species from the sample (**c**). Since antigen (Ag) binding is blocked, the result is a false-positive signal

4 Enzymes

There are several approaches to "link" enzymes as the signal reporter in ELISA. That is, conferring binding of the antigen to the immunosorbent. The direct approach is to conjugate the enzyme to the secondary or detection antibody. This can be accomplished by covalent cross-linking the enzyme to the antibody or via an SA–biotin bridge (Fig. 5). There are many methods that can be used to cross-link enzyme and antibody. Water-soluble carbodiimide chemistry such as EDAC (1-ethyl-3-{3-diethylaminopropyl} carbodimide) is useful for covalent linking of proteins. Carboxylate residues on one protein are converted into o-acyl isourea (unstable intermediate) form. In turn, the second protein couples via accessible lysine groups to form the covalent amide bond. The efficiency of the reaction can be facilitated by the addition of N-hydroxysuccinimide (NHS), which converts the EDAC into a more stable active ester [4]. There are also several protein–protein coupling kits commercially available. However, conjugates of the most popular enzymes used in ELISA, horseradish peroxidase (HRP) and alkaline phosphatase (ALP), are readily available from commercial sources.

4.1 Horseradish Peroxidase

The original peroxidase was indeed purified from horseradish. Structurally, the protein is globular containing both an α-helical region and a shorter β-sheet. An iron center (heme) group is responsible for the catalytic activity of the enzyme. There are three surface lysine residues that can be used for conjugation. The enzyme is also considered a glycoprotein with approximately 20% of the structure comprising carbohydrate. The glycan groups are thought to add to the overall stability of the enzyme. The molecular weight is about 38.8–44 kDa with an isoelectric point at pI ~5.7. There are numerous isoenzymes varying slightly in molecular

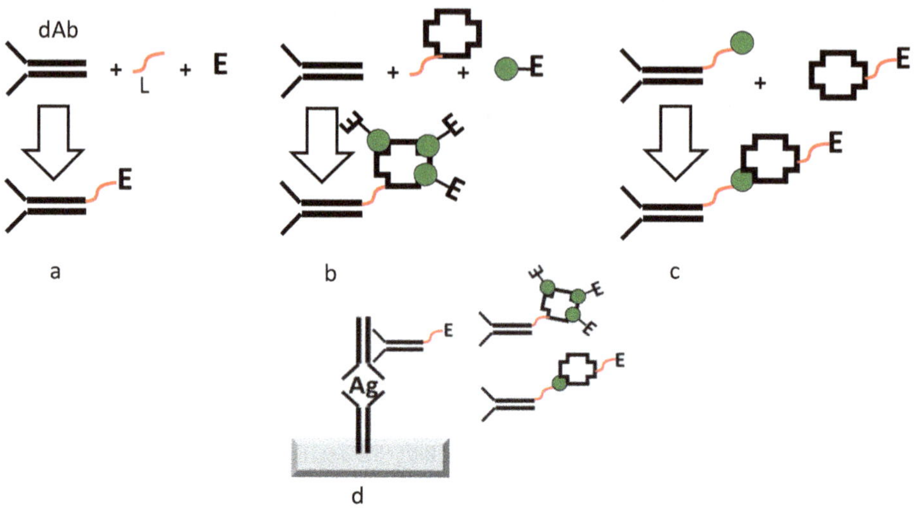

Fig. 5 Enzyme linking. There are several approaches for creating the link between enzyme (E) and the detection antibody (dAb). The direct method involves covalent linking (L) of the antibody and enzyme. For example, the antibody may be modified with amine-reactive esters. The lysine residues of the enzyme then displace the ester forming a covalent bond between enzyme and antibody (**a**). Another popular approach involves using the binding of streptavidin and biotin. Here, streptavidin is modified and linked to the antibody. Likewise, enzyme is biotinylated. Addition of the biotinylated enzyme to the streptavidin-modified antibody completes the labeling process (**b**). Alternatively, the antibody is first biotinylated, while the enzyme is linked to streptavidin (**c**). In some assays, an unlabeled detection antibody is first added. Next, an enzyme labeled anti-dAB is used to develop signal (**d**)

weight and pI. The predominant isoenzyme is HRP-C. Purity of the enzyme can be determined from the heme Soret band wavelength at absorbance A403 nm relative to the protein content measured at A280 nm. The ratio, Rz = A403/A280, is used to indicate the level of purity. HRP with an Rz value of 3.04 is considered pure. However, this purity value varies among isoenzymes and those produced from different recombinant sources [5].

The reaction mechanism for HRP is based upon cyclic redox of the heme iron. While the enzyme may act upon many different substrates, the most important catalysis is the oxidation of hydrogen peroxide that in turn can be used in coupled reaction with colorimetric, fluorescent, and chemiluminescent substrates (Table 1):

$$HRP^{\circ} + H_2O_2 \rightarrow nHRP(I) + H_20$$

$$HRP\ (I) + AH_2 \rightarrow HRP\ (II) + AH^*$$

$$HRP\ (II) + AH_2 \rightarrow HRP^{\circ} + AH^* + H_20$$

$$\text{Overall Reaction}: \quad (2)\ AH_2 + H_2O_2 \rightarrow (2)\ H_20 + (2)\ AH^*$$

Table 1
HRP signaling substrates

Substrate	Soluble	Precipitate	Colorimetric (chromogenic)	Fluorescence	Chemilum	Relative sensitivity
TMB	✓	✓	✓ Dark blue Acidified yellow			20 pg++++
4CN		✓	✓ Blue-black			5 ng+
DAB		✓	✓ Brown			1 ng++
OPD	✓		✓ Yellow-orange			70 pg+++
ABTS	✓		✓ Green			2.5 ng+
QuantaRed ADPH-based	✓			✓ 570/585		4 pg+++++
SuperSignal Luminol-based	✓				✓ 425 nm	1 pg +++++
DuoLuX Acridan-based		✓		✓	✓	

TMB 3,3′, 5,5′-tetramethylbenzidine, *4CN* 4-chloro-naphtho, *DAB* 3.3′-diaminobenzidine, *OPD* o-phenylenediamine dihydrochloride, *ABTS* 2,2′-azinobis [3-ethylbenzothiazoline-6-sulfonic acid]-diammonium salt, *ADHP* 0-acetyl-3,7-dihydroxyphenoxazine

Interference of the enzymatic activity is associated with competing substrates or the destruction of the reducing agent, AH2 [6]. Well-known inhibitors of HRP activity include EDTA, sodium azide, L-cystine, and metal cations such as Cd^{2+}, Cu^{2+}, and Fe^{3+} [7].

The most common HRP substrate for immunoassay is TMB (3,3′,5,5′ -tetramethylbenzidine). The diamine TMB has (2) aromatic amines (AH_2) from which electrons may be removed to form cation radicals (clear color) that are in equilibrium with a deprotonated (H^+) charge transfer complex (AH*). The diimine cation (yellow) adsorbs at 450 nm, while the charge transfer intermediate (blue green) absorbs at 652 nm. So, the HRP/H_2O_2 catalysis generates a blue color in the presence of TMB. Stopping the reaction with sulfuric acid shifts TMB back to the yellow-colored diimine state.

4.2 Alkaline Phosphatase

ALP [8, 9] is widely used as a conjugate for ELISA. Calf intestinal ALP is a dimeric glycoprotein of approximately 150 kDa and requires divalent cations such as zinc, magnesium, or calcium for reactivity. There are several isozymes. The pH stability of the protein is between pH 7.5 and 9.5 while the enzymatic activity is optimal between pH 8 and 10. Chelating agents are strong inhibitors of the enzyme. ALP dephosphorylates various phosphate esters, including phosphoproteins and nucleic acids. Activity is

Table 2
ALP signaling substrates

Substrate	Soluble	Precipitate	Colorimetric (chromogenic)	Fluorescence	Chemilum	Relative sensitivity
PNPP	✓		✓ Yellow			100 ng+
NBT/BCIP		✓	✓ Purple			30 pg+++
CDP-star/ CSPD Dioxetane-based		✓			✓	10 pg++++
Dynalight		✓			✓	
LumiPhos Lumigen PPD		✓			✓ 420 nm green	15 pg++++
ELF-97		✓		✓ 360/530 light green		
4-MUP	✓			✓ 372/445 blue		
AttoPhos [BBTP]	✓			✓ 435/555 yellow		
DuoLuX Acridan-based		✓		✓ 405/453	✓ 453 nm blue	

PNPP p-nitrophenyl phosphate, *NBT/BCIP* nitro blue tetrazolium/5-bromo-4-chloro-indoly phosphate, *4-MUP* 4-methylumbelliferyl phosphate, *BBTP* 2′-[2-benzothiazoyl]-6′-hydroxybenzothiazole phosphate, *Lumigen PPD* 4-methoxy-4-(3-phosphatephenyl) spiro [1,2-dioxetane-3,2′-adamantane], *ELF-97* 2-(5′-chloro-2-phosphoryloxyphenyl)-6-chloro-4(3H)-quinazolinone

usually measured in the dephosphorylation of 4-nitrophenyl phosphate (PNPP) to p-nitrophenol, which yields a yellow color. Inhibitors of ALP include chelating agents, cysteine, inorganic phosphates, and L-phenylalanine [10].

Other sources include *Escherichia coli* alkaline phosphatase (ALPK) and shrimp ALPK. The dimeric *E. coli* enzyme is not glycosylated and has a dimer molecular weight of 89 kDa. Zinc and magnesium cofactors are required for activity. Optimal activity is at pH 8. The shrimp enzyme (m. wt. 58 kDa) is less stable.

Where HRP is useful with colorimetric and chemiluminescent substrates, ALP fluorescent substrates are popular (Table 2). The substrate AttoPhos produces a highly fluorescent soluble product for high-sensitivity detection. Others include 4-MUP and DuoLuX, which also has a chemiluminescent signature. See Tables 1 and 2 for an expanded (but not exhaustive) list of enzyme substrates. Keep in mind that not all substrates are suitable for a particular application. For example, certain substrates that were designed for use with Western blot membranes such as nitrocellulose or PVDF may not be suitable for plastic well-based assays. Conversely, soluble substrates may not adsorb properly to membrane-based ELISAs yielding high backgrounds or nonuniform signal development. There are also enzymes other than HRP or ALP such as glucose oxidase, beta-galactosidase, or catalase that may be used under certain conditions.

5 Streptavidin–Biotin Systems

SA is a naturally occurring biotin-binding protein originally isolated from the Gram-positive, filamentous bacterium *Streptomyces avidinii* [11]. The endogenous functionality of SA regarding the bacteria is of no concern, but rather its profound utility in the linking together of dissimilar biomolecules. The SA–biotin binding (Ka ~ 10^{15}) is thought to be the strongest noncovalent chemical bond in nature. Moreover, the tetrameric SA can bind up to 4 moles of the vitamin per mole of protein. Thus, enzymes such as HRP and ALP are commonly cross-linked to SA for use as signaling agents while leaving biotin-binding domains available for binding of a biotinylated antibody—the essence of ELISA. This has been unquestionably a well-accepted approach until most recently when vitamin B7 (biotin) food supplements began to be used in significant quantity. This has led to a carryover of high levels of free biotin into the body fluids. An immunoassay built using the SA–biotin reporter system is thus potentially subject to biotin interference [12, 13]. For those assays in which an SA particle is used to sequester a biotinylated capture antibody, the excess of free biotin can lead to competitive binding assay interference (Fig. 6).

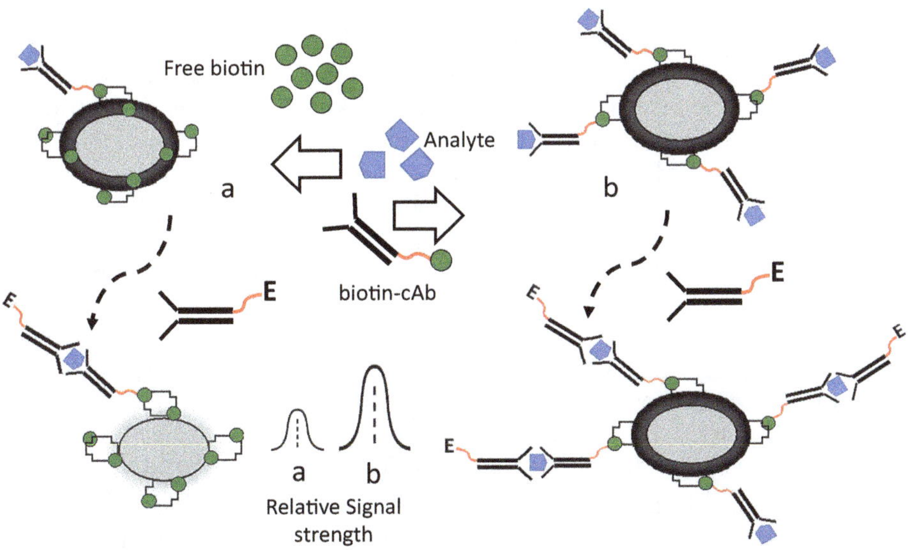

Fig. 6 Biotin interference. The presence of free biotin in the sample becomes problematic when it competes with the biotinylated cAb for binding sites on the streptavidin particle. In effect, this reduces the signal associated with analyte

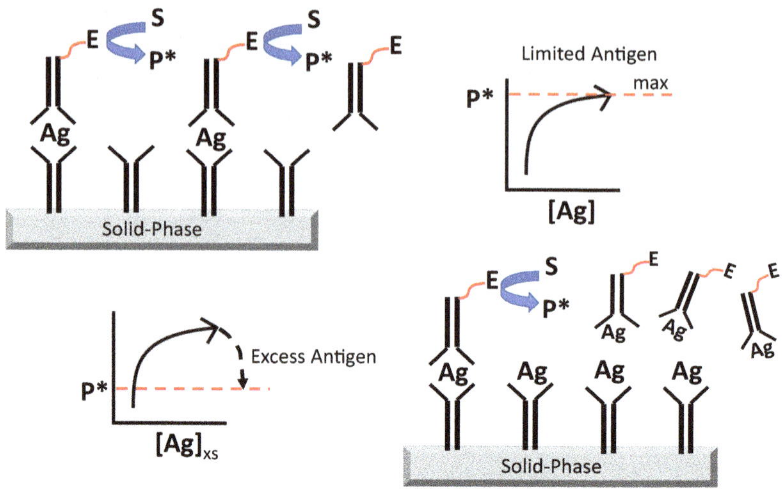

Fig. 7 The hook effect. Under limited antigen (Ag) conditions usually encountered by serial dilution of the sample, the dose–response curve shows an increased signal (P) with increased dose plateauing to a maximum value (P*). However, if excessive amounts of antigen are present in the sample, it is possible to saturate both binding of the solid phase capture antibody and secondary reporter antibody. In this case, the signal can be significantly reduced to a lower value at high dose, often resulting in a false-negative report

6 The Hook Effect

At high levels of antigen or enzyme substrate, it is possible to achieve a false-negative result from the assay. The primary effect is due to binding saturation. For an enzyme reaction, this means that the substrate is present in such great excess that the formation of product in excess literally inhibits or retards further conversion of substrate to product. This is known as substrate–product inhibition. Similarly, the hook effect is often associated with a sandwich immunoassay in which antigen is in excess. Under high antigen dose, the capture antibody-binding sites, as well as detection antibody, are fully occupied or saturated with bound antigen [14]. The result is a reduction in dose–response at higher levels of antigen (Fig. 7).

7 The Importance of Buffers

The choice of buffer(s) to be used in an ELISA is critical to attain optimal performance. Buffers are used to block surfaces, others to stabilize antibodies and enzymes, still others to clear the surface of nonspecific binders. Some contain proteins, surfactants, rehydrating agents, sugars, lipidic or synthetic polymers, along with antimicrobials. How these interact with surfaces, antibodies, and enzymes requires evaluation in order to achieve the desired outcome for the assay. Moreover, the biological matrix (e.g., serum, plasma, urine) is most likely to influence the choice of buffers. The role of buffers in building a functional ELISA is explained in the following sections.

7.1 Coating Buffer

After selecting a suitable immunosorbent substrate (e.g., microplate), the next step is to prepare the capture antibody for immobilization to the surface in a suitable buffer (the coating buffer). The coating buffer must be compatible with the antibody stability to avoid protein denaturation, unwanted precipitation, and fowling of the dispenser used in delivery of the antibody to the microplate. The buffer should be absent of any additives that would adversely modify the surface or compete against the antibody immobilization. Typical buffers for this purpose are phosphate-buffered saline (PBS), pH 7.4, sodium carbonate–bicarbonate, 50–150 mM, pH 9, or sodium borate, 50 mM, pH 9. For bulk immobilization (coating) by dispense of the antibody solution to the whole well (e.g., 100 μL), the stock antibody is diluted to about 1–10 μg/mL in a coating buffer. If antibody solution is to be micro-dispensed (e.g., microarray), then the input concentrations are generally higher on the order of 0.2–1 mg/mL when printing at delivery volumes of picoliter (pL) to nanoliter (nL).

If protein solubility is a problem, it may be an issue with improper selection of the pH. If you know the pH isoelectric point (pI) of the protein (net zero charge), adjust the buffer pH upward (more basic) to increase solubility. Proteins are generally less soluble at their pI and are more prone to precipitate under acidic conditions (pH 4–6) rather than basic pH 7–9. If solubility of the antibody protein does not improve, try serial dilutions in buffer followed by coating with the dilutions in wells overnight under refrigeration. Check for coating efficiency, for example, by using a labeled antispecies secondary antibody. If dilution or pH adjustments do not improve solubility, the addition of detergent additive such as Tween-20 is recommended.

7.2 Wash Buffer

Once antibody coating of the well is accomplished, it is advisable to first wash away any loosely bound protein using a detergent-based buffer such as PBST (phosphate-buffered saline containing 0.05–0.1% Tween-20) or TBST (Tris-buffered saline + Tween 20). The surfactant Tween-20 may serve two roles: removal of the nonspecific protein while maintaining hydration of the bound protein to reduce denaturation.

7.3 Blocking Buffer

After coating with antibody (or antigen for indirect ELISA), it is critical to block the surface to reduce nonspecific binding that may lead to high signal backgrounds. This is because most coatings do not lead to uniform and full coverage of the surface. Washing steps may also strip away a portion of the bound antibody exposing the surface. As a result, the cleared surface must be blocked.

There are many kinds of blocking buffers. It is advisable to screen several of these against the biological matrix to be used in the assay in order to rule out adverse effects such as high backgrounds or interferences. Typically buffers employed for blocking will contain an additive that adsorbs to the exposed surface. For example, PBS containing 1% BSA, pH 7,4 or carbonate buffer, pH 9 containing 1% casein. There are also additives that may be included in order to reduce interference within the matrix such as the presence of heterophilic antibodies.

7.4 Diluents

7.4.1 Antibody Diluent

While coating and blocking are necessary steps in preparing the capture plate, it is also important to stabilize the assay development reagents, especially the secondary or reporting antibody and the signaling reagents (enzyme and enzyme substrate). Buffers used in the blocking step may be suitable as antibody diluents. Common diluents include PBS–1% BSA, pH 7.4, PBS–0.05% Tween 20 (PBST), or Tris-buffered saline (TBST), as well as HEPES buffer (50 mM HEPES, 0.1 M NaCl, 1% BSA, pH 7.4). Carbonate buffers at pH 8.5–9.5 with BSA, casein, or human serum albumin (HSA) are also used. There are available nonprotein blocking agents if BSA, HSA, or casein are not suitable.

7.4.2 Matrix Diluent

An often-encountered fallacy is to rely upon the creation of dose–response curves using analytical standards or controls diluted in a simple buffer to estimate analyte (antigen) levels from a biological sample (matrix). This can potentially lead to an underestimate (false negative) from interfering substances in the sample (but not the standard) that suppress signal by some means such as interfering with specific binding between antibody and analyte. On the other hand, an overestimate (false positive) may be created by the presence of heterophilic antibody in the sample.

The simplest approach to overcoming sample matrix effects is to dilute. This may aid in dissociation of an interfering substance to an acceptable level or reduce the viscosity of the sample, thereby improving analyte transport kinetics for more effective binding leading to enhanced signaling. In addition to the common diluents described earlier, there may be the need for additional additives to overcome specific matrix effects in the sample due to, for example, divalent ion concentration and other interfering substances such as carbohydrates, lipids, glycoproteins, etc. Proprietary additives that reduce heterophilic antibody interference are commercially available.

7.4.3 Checking for Matrix Effects

The easiest assessment is to perform a spike and recovery assay in which a known amount of analyte standard is added (spiked) into a sample. The spiked sample is then compared to analyte standard diluted into buffer based upon the dose–response curve. Acceptable levels of recovery (the expected response) from standard dose are about 80–120%. Outside of that range, a matrix effect is suggested.

References

1. Gibbs J, Vessels M, Rothenberg M (2017) Immobilization principles – selecting the surface for ELISA assays. Corning Incorporated, Life Sciences, Application Note. https://www.corning.com/catalog/cls/documents/application-notes/CLS-DD-AN-454.pdf
2. Bolstad N, Warren DJ, Nustad K (2013) Heterophilic antibody interference in immunometric assays. Best Pract Res Clin Endocrinol Metab 27:647–661
3. Juncker D, Bergeron S, Laforte V et al (2014) Cross-reactivity in antibody microarrays and multiplexed sandwich assays: shedding light on the dark side of multiplexing. Curr Opin Chem Biol 18:29–37
4. Matson RS (2021) Chapter 2: well-based antibody arrays. In: Whitaker KC, Huang R-P (eds) Antibody arrays: methods in molecular biology, 2237: 11–38, Springer protocols. Humana Press, New York
5. Krainer FW, Glieder A (2015) An updated view on horseradish peroxidases: recombinant production and biotechnological applications. Appl Microbiol Biotechnol 99:1611–1625
6. Josephy PD, Eling T, Mason RP (1982) The horseradish peroxidase-catalyzed oxidation of 3,5,3′,5′-tetramethylbenzidine. Free radical and charge-transfer complex intermediates. J Biol Chem 257(7):3669–3675
7. https://www.sigmaaldrich.com/US/en/technical-documents/technical-article/protein-biology/enzyme-activity-assays/peroxidase-enzymes
8. Luis Millan JL (2006) Alkaline phosphatases-structure, substrate specificity and functional relatedness to other members of a large superfamily of enzymes. Purinergic Signal 2:335–341. https://doi.org/10.1007/s11302-005-5435-6
9. Sharma U, Pal D, Prasad R (2014) Alkaline phosphatase: an overview. Ind J Clin Biochem 29(3):269–278. https://doi.org/10.1007/s12291-013-0408-y
10. Phosphatase, Alkaline from bovine intestinal mucosa https://www.sigmaaldrich.com/US/en/product/sigma/p5521
11. Tausig F, Wolf FJ (1964) Streptavidin–a substance with avidin-like properties produced by microorganisms. Biochem Biophys Res Commun 14(3):205–209
12. Avery G (2019) Biotin interference in immunoassay: a review for the laboratory scientist. Ann Clin Biochem 56(4):424–430
13. Li D, Ferguson A, Cervinski MA et al (2020) AACC guidance document on biotin interference in laboratory tests. [Epub]. J Appl Lab Med 5:575. https://doi.org/10.1373/10.1093/jalm/jfz010
14. Sequeira S (2019) An overview on interference in clinical immunoassays: a cause for concern. Hamdan Med J 12(4):158–164

Chapter 3

Antibody Immobilization

Robert S. Matson

Abstract

In the ELISA format(s), the capture antibody is usually affixed to a solid phase, commonly referred to as the immunosorbent. How to tether the antibody most effectively will depend upon the physical properties of the support (plate well, latex bead, flow cell, etc.) as well as its chemical nature (hydrophobic, hydrophilic, the presence of reactive groups such as epoxide, etc.). Of course, it is ultimately the suitability of the antibody to withstand the linking process while preserving its antigen-binding efficiency that must be determined. In this chapter, the antibody immobilization processes and their consequences are described.

Key words Antibody, Immobilization, Solid phase, Coupling, Chemistry, Oriented, Random, Passive, Density, Steric hindrance, Protein A, Streptavidin, Biotin

1 Introduction

In this chapter, we discuss the various approaches and methods leading to successful immobilization of antibody to a solid support. The chapter begins with a description of the solid support surface chemistries and strategies commonly used for immobilization of antibodies. This is followed by guidance on how to prepare the antibody for surface attachment.

1.1 Surface Chemistries

We have already discussed the more general aspects surrounding the preparation of ELISA (see chapter on ELISA essentials) such as passive adsorption of antibody to the support. Thus, we recognize that medium- to high-binding plastic surfaces (e.g., polystyrene microplate) that are specifically designed for antibody immobilization are indeed the most common approach for creating an ELISA. The passive coating of antibody (or antigen) to microplate wells is the easiest and most straightforward approach. It should be tried first! Usually, after a few attempts at optimizing the coating concentrations, incubation conditions, and buffers, the process leads to a successful immunoassay. If, on the other hand, the assay appears

Robert S. Matson (ed.), *ELISA: Methods and Protocols*, Methods in Molecular Biology, vol. 2612,
https://doi.org/10.1007/978-1-0716-2903-1_3,

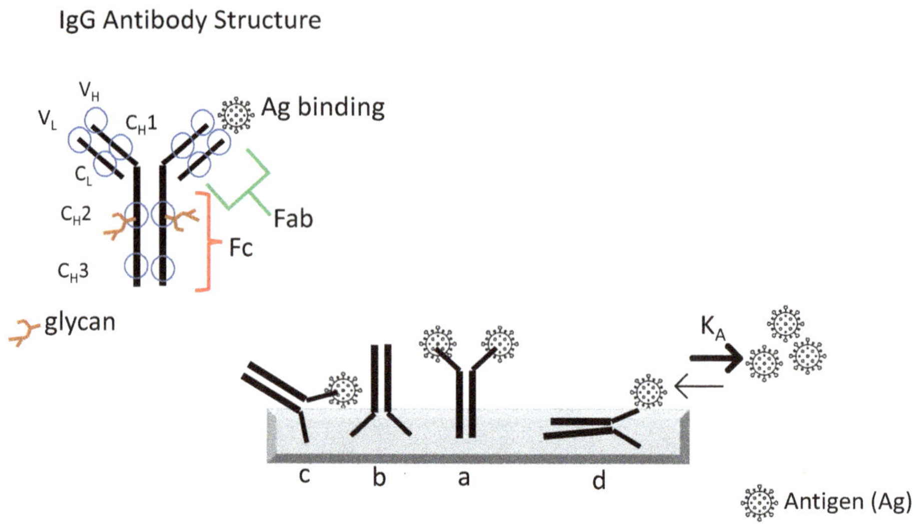

Fig. 1 Passive adsorption. Antibodies are randomly associated with the solid support's surface. This can result in the molecule sequestered via its Fc region (**a**) allowing full antigen-binding capacity (avidity) or conversely adsorption via the Fab- binding site regions (**b**), thereby preventing antigen capture or tethered to the support partially through the binding site (**c**) reducing but not fully blocking antigen binding. In the most extreme state (**d**), the antibody is strongly adsorbed and partially denatured, thereby reducing the antibody's binding affinity

to yield lower or unacceptable levels of sensitivity and/or specificity, it may be necessary to consider other approaches.

First, let us consider how passive adsorption may have failed. What factors influence the outcome? Is it surface reactivity-related or a problem with the antibody? Passive adsorption often leads to a random oriented (or stochastic orienting) of the antibody onto the surface (Fig. 1). That is, the antibody may be tethered to the surface in any number of orientations. For example, if the antibody is partially immobilized via one of the antigen-binding regions (facing the surface), then the antibody's avidity is compromised. For IgG, one binding site is open while the other is sterically hindered by the surface. If the antibody when surface bound is partially denatured, then the binding affinity is likely to be reduced as well assuming the binding sites have been destroyed or altered. So, the ideal orientation is for the antibody-binding sites to be facing away from the surface and fully available for antigen capture (Fig. 2). It has been reported in earlier work that antibody adsorbed onto polystyrene in a random fashion exhibits only about 5–10% of the native-binding capacity, while antibody linked via biotin–streptavidin retained about 70% of the native activity [1]. Kausaite-Minkstimiene et al. [2] also found that the oriented coupling of anti-HGH antibody to an SPR-chip provided 8.9–11.2-fold greater antigen binding than that of the randomly adsorbed antibody. The loading on the randomly adsorbed antibody was about twofold higher than that of the oriented antibody. However, there are other factors such as pH

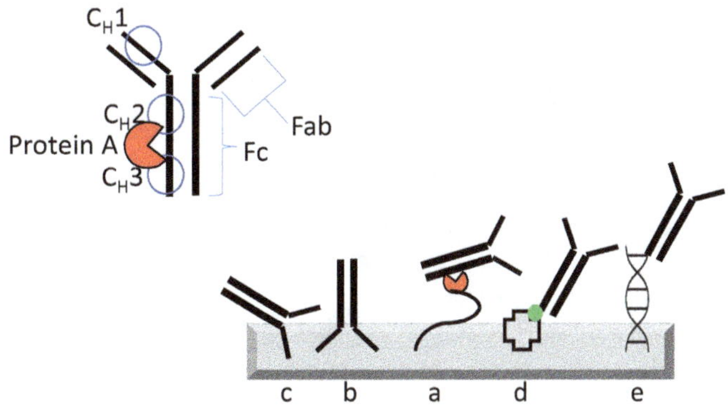

Fig. 2 Random versus oriented tethering. Random orientations may lead to blocking of a single binding site (**c**) or both binding sites for IgG (**b**). Oriented tethering can be accomplished by using Fc-specific affinity ligands (**a**). IgG-binding proteins such as Protein A and Protein G are often used for this purpose [13]. Applying a coat of streptavidin to the surface has been successfully used to increase binding activity of biotinylated antibodies (**d**). DNA-directed immobilization of oligonucleotide–antibody conjugates has been demonstrated to improve immunoassay sensitivity (**e**)

(ionic charge), molecular weight (size), and hydrophobicity that may influence the outcome of immobilization. Matson and Little [3] first observed that the coupling of antibodies to a chromatographic porous bead support in a manner that provided optimal antigen-binding affinity was most favorable at lower antibody density and tethering via carbohydrate localized to the Fc region. Susini et al. [4] compared antibodies that were immobilized either by random or oriented means to polystyrene plates. They examined the impact on antigen-binding capacity in relationship to the size and valency (number of available epitopes) of the protein antigen. Antibodies directed to small monomeric antigens (troponin, 24 kDa, 1 epitope) that were oriented outperformed those randomly coupled. With larger, monovalent antigens (PSA-ACT, 90 kDa, 1 epitope), there was no significant difference in antigen binding. However, larger, multivalent protein antigen (ferritin, 474 kDa, 12 epitopes) binding was greater using randomly coupled antibody. So, smaller antigen binding was optimal when antibody was oriented with all binding sites available and free of steric hindrance (Figs. 3 and 4a). As the antigen size increased, steric hindrance prevented efficient binding to the antibody (Fig. 4b). However, with large proteins that exhibited multiple copies of the epitope it was possible to increase the number of binding events (Fig. 4c). Thus, the larger population of adsorbed (random) antibody presumably contributed more total binding sites than oriented antibody.

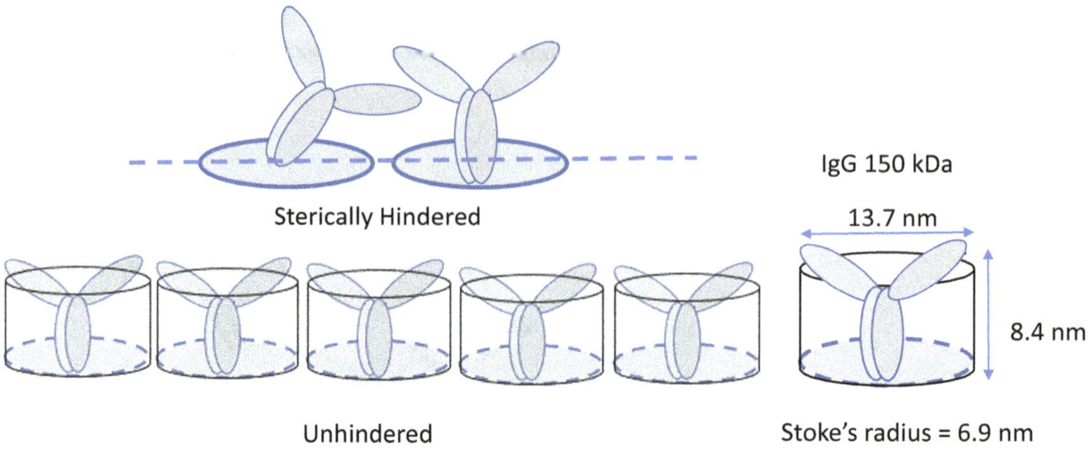

Fig. 3 Surface packing density and steric hindrance (SH). Immobilization of antibodies and other proteins onto surfaces is limited by surface charge, pH, hydrophilic or hydrophobic properties, attachment strategy (random versus oriented), molecular size, and steric hindrance [17]. Excessive surface loading (crowding) leads to steric hindrance (SH), which can greatly reduce the binding efficiency. In general, binding proteins perform optimally at surface distribution greater than their Stoke's radius

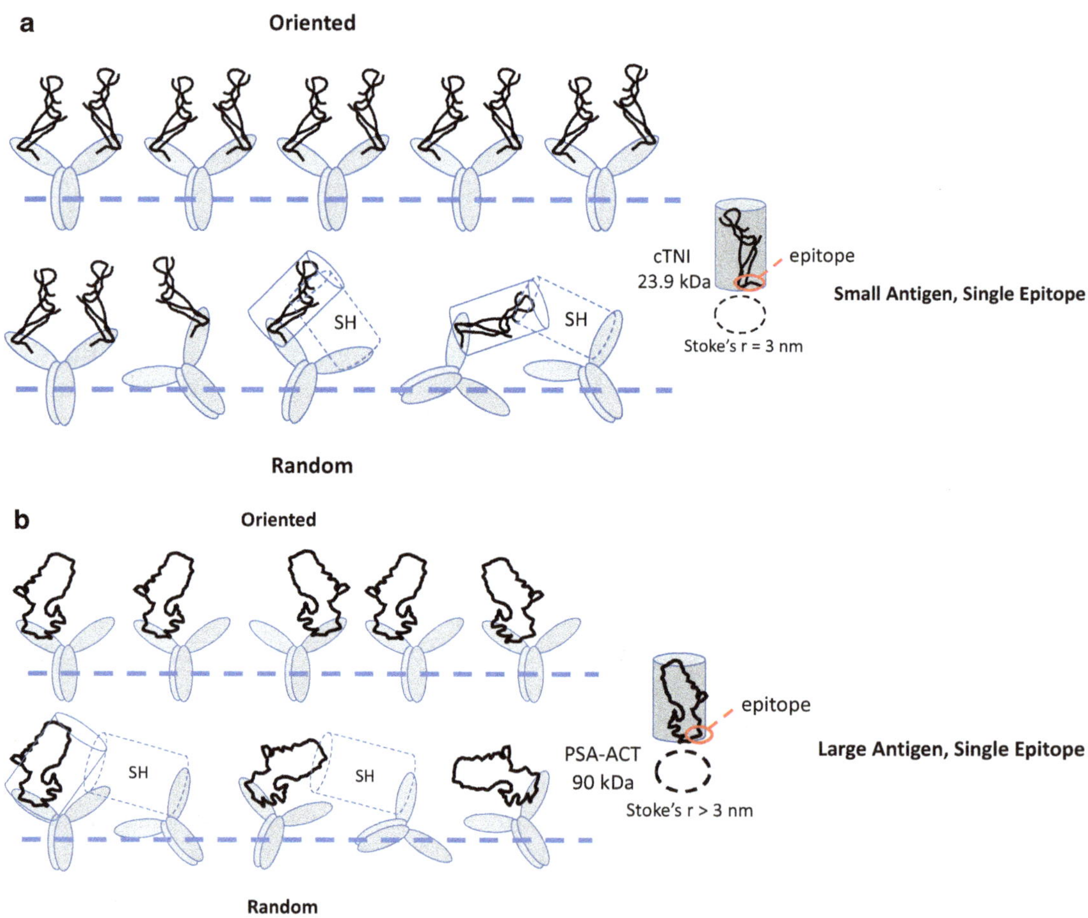

Fig. 4 Random versus oriented tethering for (**a**) a small antigen with single epitope for antibody binding, (**b**) a large antigen with single epitope for antibody binding, and (**c**) large antigen with multiple epitopes for antibody binding

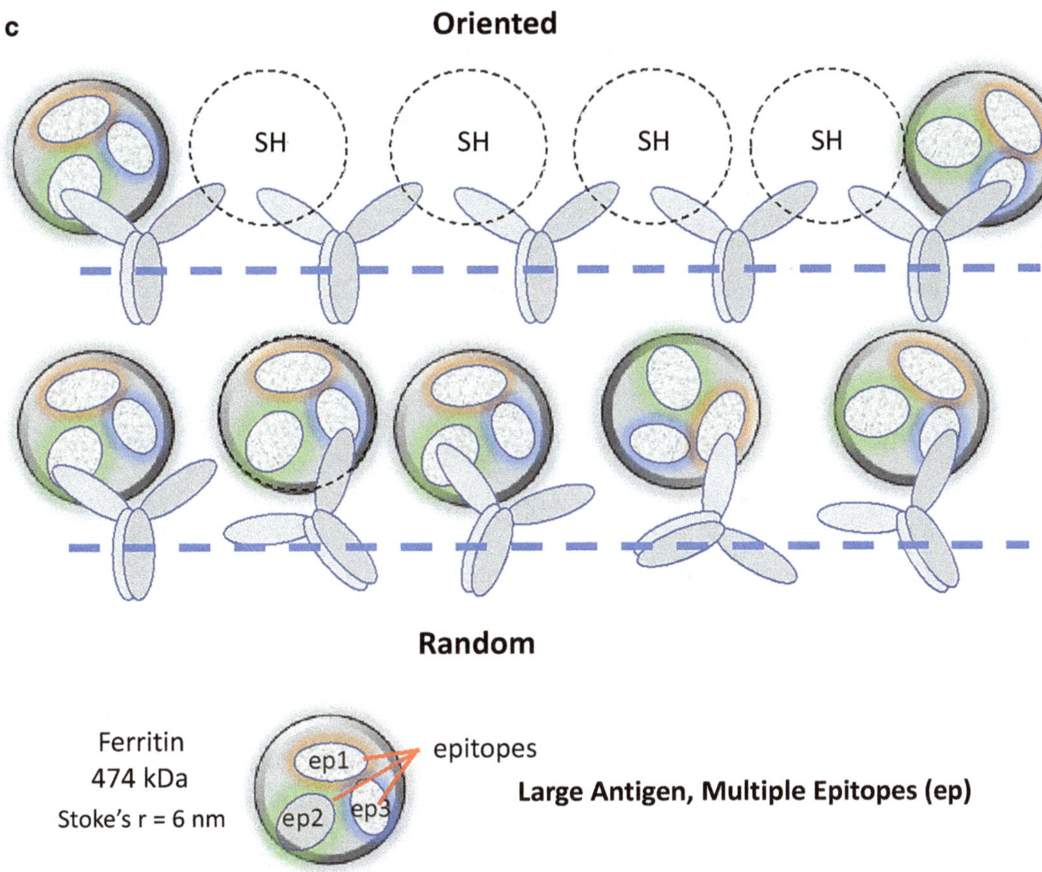

Fig. 4 (continued)

1.2 Coupling Strategy

Sonawane and Nimse [5] provide a review of various surface materials (solid supports: silicon, glass, carbon, nitrocellulose, gold, silver, polystyrene) and attachment chemistries (physical adsorption and covalent: aldehyde, epoxy, EDC/NHS carboxylation, hydrazide, amination) useful for protein and nucleic acid-based diagnostic devices. Welch et al. [6] examined various surface modification of the solid support that promotes antibody-antigen binding and techniques useful in the characterization of the bound antibody.

A detailed review of coupling chemistries available for oriented immobilization of antibodies has recently been provided [7].

An interesting "one-step" approach involves the formation of solution phase complex of 3-aminopropyltriethoxysilane (APTES) with antibody [8]. The APTES-antibody complex apparently increases both the ionic and hydrophobicity of the antibody, thereby promoting rapid adsorption to the polystyrene surface. Results were compared with covalent coupling and conventional adsorption. The limits of detection (LOD, pg/mL) were one-step (7), covalent (12), and passive adsorption (226), while the antibody

Table 1
Comparison of commercial microplate surfaces for ELISA

Mfg/vendor	Brand/type	Polystyrene plate surface chemistry	Binding capacity (ng IgG/cm^2)
Nunc	*MaxiSorp* (HB)	Adsorption	600–650
Pierce	*Corner Notch* (HB)	Adsorption	400 [a](128 ng/well)
Greiner Bio-one	*Microlon 600* (HB) *Microlon 200* (MB)	Adsorption	600 [a](192) 200 [a](64)
Corning	High Bind	Adsorption	400–500
Pierce	Protein A	Affinity: Fc-IgG	2–5 pmol/well
G-Biosciences	Protein A	Affinity: Fc-IgG	4 pmol/well
Biomat	Protein A, Protein G	Affinity: Fc-IgG	2.7–3.3 pmol/well 400–500 ng/well
Sigma	*SigmaScreen* streptavidin High capacity	Affinity: biotin	>300 pmol/well biotin
G-Biosciences	*Well-Coated* streptavidin	Affinity: biotin	15 pmol/well biotin
Acro Biosystems	Streptavidin	Affinity: biotin	5 pmol/well biotin

HB high binding, *MB* medium binding
[a]Surface area of a 6.4-mm-diameter well is approximately 0.32 cm^2. IgG (150 kDa) ~ 150 ng/pmol

immobilization time (hours) was one-step (0.5), covalent (2.5), and passive adsorption (14).

Thus, as discussed, we find that there are many surface chemistries and methodologies useful for immobilization of antibodies on a variety of solid-phase supports from which to develop ELISA. The problem is that only a few are commercially available (*see* Table 1). For the development of ELISA using the passive adsorption (coating) of antibody onto polystyrene plates, manufacturers have modified the surface for IgG-binding capacity (ng IgG/cm^2). ELISA plates are thus categorized as medium- to high-binding capacity. A medium-binding plate (untreated, hydrophobic) with a capacity of 100–200 ng IgG/cm^2 is designed for the capture of larger, more hydrophobic antibodies, while a high-binding plate (carboxyl surface) provides both ionic and hydrophobic interactions. The high-binding (400–600 ng IgG/cm^2) surface is generally preferred since IgG loading is higher, allowing better sensitivity in detection of lower molecular weight antigens [9]. The surface-bound antibody antigen-binding capacity may also be affected depending upon the extent of glycosylation, the molecular structure of the glycan chain (s), and attachment site(s) within the immunoglobulin [10, 11]. It is therefore prudent to evaluate different surfaces. Passive adsorption is also affected by pH conditions. Polystyrene is largely hydrophobic, and proteins adsorb to its surface primarily via hydrophobic interaction. Shifting the pH toward the apparent isoelectric point (pI) of the protein may allow unfolding, resulting in exposure of previously buried hydrophobic regions of the protein. This conformational change facilitates the binding of the protein to polystyrene's hydrophobic surface. Human IgG (monoclonal) has a pI ~6.6–7.2, but subtypes such as IgG_1 are reported at pI 8.6 ± 0.4 [12]. Adjusting toward a basic pH > pI is recommended for passive adsorption. For IgG, try adjusting to pH 9–10 for coating.

If an oriented presentation of the capture antibody is desired, then Protein A- or Protein G-coated surfaces are recommended. These bind specifically to the Fc region of IgG. There are notable differences in binding efficiency among IgG subclasses to consider. For example, Protein A fails to bind human IgG_3, while Protein G shows high affinity for the subclass [13]. Another approach is to utilize a streptavidin coating for tethering of a biotinylated antibody. The Fc region of antibodies is glycosylated [14] and can be chemically modified with biotin-hydrazide [15]. Thus, the antibody is oriented through the streptavidin–biotin binding. This biotin-hydrazide reagent is commercially available from numerous sources. However, a word of caution—glycosylation of the antigen-binding Fab region has been reported to occur in 15–25% of IgG [16].

If necessary, antibodies may be immobilized by covalent means to the support (*see* Table 2). For example, if the assay requires

Table 2
Commercially available covalent cross-linking microplates for ELISA

<table>
<tr><th>Mfg/Vendor</th><th>Brand/Type</th><th>Plate |--R group Surface Chemistry</th><th>Coupling Chemistry</th><th>Surface Attachment</th></tr>
<tr><td>Nunc</td><td>CovaLink NH
2º Amine</td><td>|--CH$_2$-NH: (CH$_3$) →
Shift from pH 5 to pH 7.2</td><td>IgG-COOH + EDC/NHS →
IgG-CONHS →
Carbodiimide reaction</td><td>|--CH2-N(CH$_3$)-C(=O)-IgG</td></tr>
<tr><td>Biomat</td><td>Carboxylated</td><td>|--COOH +
EDC/NHS → |--CONHS
Carbodiimide reaction</td><td>+ IgG-NH: →
Shift from pH 5 to pH 7.2</td><td>|--C(=O)-NH-IgG</td></tr>
<tr><td>PolyAn</td><td>3D Epoxy</td><td>|--CH2-CH-CH (epoxide, O)</td><td>+ IgG-NH: →
pH 9-10
+ IgG-SH: →
pH 7.5-8.5</td><td>|--CH2-CH(OH)-CH — NH-IgG
S-IgG</td></tr>
<tr><td>Pierce</td><td>Maleimide Activated</td><td>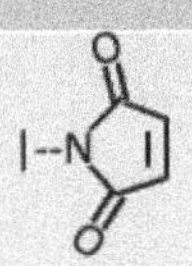</td><td>+ IgG-SH: →
Reduce antibody to break disulfide (S-S) bonds
pH 6.5-7.5</td><td>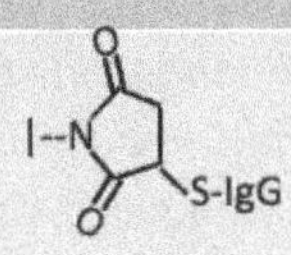</td></tr>
</table>

rigorous or harsh rinsing of the surface to remove nonspecific binders with additives (such as SDS or urea) that would otherwise strip away antibody, then covalent attachment may help preserve the antibody loading. Both aminated (-NH_2) and carboxyl (-COOH) microplates are commercially available for use. These are stable surfaces that can be modified with reactive esters using carbodiimide chemistry. Once the ester is formed, then covalent coupling may occur rapidly with the introduction of a primary amine. In the case of aminated surface, it would react with an antibody that had been modified (esterified) using the carbodiimide (EDC/NHS) process. Conversely, the carboxyl surface would be first esterified, and the unmodified antibody coupled to the support via lysine residues. These reactions occur at neural pH 7.2 in buffer free of amine or sulfhydryl, which would unnecessarily compete with the protein-coupling reaction.

As noted earlier, depending upon the pI of the antibody, it may be necessary to employ a buffer at higher pH. Supports containing surface epoxides are more reactive toward proteins under basic conditions, pH 9–10 a very stable linkage. Reduced antibodies with thiol (-SH) groups can also be coupled via epoxide at pH 7.5–8.5. Another approach for immobilization of reduced antibody involves reaction with support containing surface maleimides.

The following general protocols are provided as a starting point in the preparation of the ELISA plate using passive adsorption of unmodified antibody (random coupling), biotinylated antibody binding to streptavidin, or Protein A capture of antibody onto polystyrene plates for oriented couplings. For covalent immobilization, it is recommended that the vendor's instructions be closely followed since their protocols are optimized for their product performance.

2 Materials

2.1 Passive Adsorption of Antibody to Polystyrene Plate

1. Solid phase: polystyrene plate, 96-well, ELISA grade medium or high binding.
2. PBS, pH 7.4 coating buffer (*see* **Note 1**).
3. TBST: Tris-buffered saline with 0.05% (v/v) Tween-20.
4. BSA blocker: 1% bovine serum albumin, fraction V in PBS, pH 7.4 (*see* **Note 2**).

2.2 Immobilization of Biotinylated Antibody to Streptavidin-Coated Plate

1. Streptavidin-coated microplate.
2. Biotinylated capture antibody (*see* **Note 3**).
3. TBST wash buffer.
4. TBST containing 1% BSA.

2.3 Protein a Plate Capture of Monoclonal Antibody

1. Protein A-coated microplate.
2. Capture antibody subclass: mouse monoclonals IgG2a, IgG2b, IgG3 *but not* IgG1 (*see* **Note 4**).
3. TBST wash buffer.
4. BSA blocker.

3 Methods

3.1 Passive Adsorption of Antibody

This protocol is for the preparation of capture antibody by direct adsorption to polystyrene ELISA grade microtiter plates.

1. Capture antibody is prepared at 1 μg/mL in PBS, pH 7.4, and applied at 100 μL per well (*see* **Note 5**).
2. The plate wells are sealed and incubated overnight at room temperature.
3. The wells are rinsed 3 × 200 μL/well in TBST.
4. The plate is air-dried, sealed, and stored under refrigeration prior to use.
5. For blocking, bring the plate to room temperature.
6. Prepare BSA blocker and apply 200 μL/well for 1 hour at room temperature (*see* **Note 2**).
7. Rinse wells in TBST as previously described in **step 3**.
8. Invert the plate over a clean adsorbent (napkin) and air-dry overnight at room temperature (*see* **Note 6**).
9. Seal the plate and store dry under refrigeration prior to use.

3.2 Immobilization of Biotinylated Antibody

This protocol is for immobilization of a biotinylated capture antibody by binding to a streptavidin-coated polystyrene microtiter plate.

1. Bring the streptavidin-coated microplate to room temperature.
2. Rinse 3 × 200 μL/well in TBST.
3. Biotinylated capture antibody is prepared at 10 μg/mL in TBST–1% BSA and applied at 100 μL per well (*see* **Note 7**).
4. Incubate the plate for 1–2 h with shaking at room temperature.
5. Rinse wells 3 × 200 μL/well in TBST.
6. The plate is air-dried, sealed, and stored under refrigeration prior to use.

3.3 Protein a Capture of Monoclonal Antibody

This protocol is for sequestering of an IgG antibody by Protein A-coated microtiter plate.

1. Bring the Protein A-coated microplate to room temperature.
2. Wash wells 3 × 200 μL TBST.

3. Prepare the antibody at 1 μg/mL in BSA blocking buffer (*see* **Note 8**).
4. Apply diluted antibody at 100 μL/well.
5. Incubate the plate for 1 h at room temperature with shaking.
6. Rinse wells 3 × 200 μL TBST.

4 Notes

1. Coating buffer: Carbonate buffer, pH 9.6, or borate buffer, pH 9.3, is also used.
2. Blockers: While BSA and nonfat dry milk (casein) are commonly used as blocking agents in coating buffers, on occasion nonspecific binding to various grades have been reported. It is advisable that other blocking agents and coating buffers be tested for verification of optimal performance for a particular ELISA microplate. Several vendors now offer BSA-free blockers as well as nonprotein blockers as replacements.
3. Biotinylated capture antibody: If possible, source vendors who offer biotinylated antibodies (used as secondary antibody reporters) that can be evaluated as a capture antibody for the ELISA. Otherwise, purchase commercially available biotinylation kit(s) to prepare small batches using purified antibody. It will be necessary to remove excess, free biotin from the preparation. There are spin-columns to remove free biotin (Zebra brand, Pierce Thermo Fisher), which is also included in their biotinylation kit. Free biotin can also be removed by dialysis. Vector Labs offers the QuanTag kit for quantitation of free biotin in a sample.
4. Capture antibody species: The relative affinity and selectivity of immunoglobulin binders such as Protein A, G, and L toward antibody subtypes and species are well known and should be reviewed prior to using coated plates. Sigma's Product Information sheet for Protein A-coated plates (PN: S 1938) contains Table 2: Protein A, G, and L Capacities for Various Species.
5. Loading conditions: IgG adsorption capacity of high-bind polystyrene plates is approximately 100–200 ng/well. Applying 100 μL of 1 μg/mL IgG provides about 100 ng IgG/well. Depending upon incubation time, agitation, and temperature, the amount adsorbed will vary. Adjust the input IgG concentration (e.g., 0.05–10 μg/mL) to achieve efficient adsorption and ELISA sensitivity.
6. Storage conditions: The prepared plate may also be stored dry, sealed at 2–8 °C. If the antibody activity is diminished, then the addition of immunoassay stabilizers may be required to

maintain the antibody's native conformation for optimal antigen binding. Surmodics (www.surmodics.com) offers StabilBlock, StabilCoat, and StabiGuard products for this purpose. Other vendors include Meridian Bioscience (www.meridianbioscience.com/lifescience) and Candor (www.candor-bioscience.de).

7. Input concentration of biotinylated antibody: 100 μL of 10 μg/mL biotin-IgG is estimated to deliver ~100 pmol biotin per well. [Calculation: biotin (m. wt. 244 g/mol or ng/nmol) conjugation is reported to be in the range of 2–30 nmol/nmol protein (Vector Labs). For IgG (150 kDa or 150,000 g/mol or ng/nmol) and biotin conjugated at 15 nmol/nmol IgG, the conjugation would yield about 0.0244 ng biotin/ng IgG or 24.4 ng/μg IgG or ~100 pmol biotin/μg IgG. Thus, 100 μL of 10 μg/mL biotin-IgG delivers 100 pmol biotin per well.]
8. Input concentration of IgG: Delivery of 100 μL of 1 μg/mL antibody is about 100 ng IgG/well or about 0.67 pmol/well. The Protein A-binding capacity (*see* Table 1) is 2–5 pmol/well. Adjust the input IgG concentration to achieve the desired sensitivity.

References

1. Butler JE, Ni L, Nessler R, Joshi KS, Suter M et al (1992) The physical and functional behavior of capture antibodies adsorbed on polystyrene. J Immunol Methods 150:77–90
2. Kausaite-Minkstimiene A, Ramanaviciene A, Kirlyte J, Ramanavicius A (2010) Comparative study of random and oriented antibody immobilization techniques on the binding capacity of Immunosensor. Anal Chem 82:6401–6408
3. Matson RS, Little MC (1988) Strategy for the immobilization of monoclonal antibodies on solid phase supports. J Chromatogr 458:67–77
4. Susini V, Fierabracci V, Barria G, Dodoli L et al (2021) Signal enhancement in oriented immunosorbent assays: a balance between accessibility of antigen binding sites and avidity. Biosensor 11:1–10. https://doi.org/10.3390/bios11120493
5. Sonawane MD, Nimse SB (2016) Surface modification chemistries of materials used in diagnostic platforms with biomolecules. J Chem 2016:1–19. https://doi.org/10.1155/2016/9241378
6. Welch NG, Scoble JA, Muir BW et al (2017) Orientation and characterization of immobilized antibodies for improved immunoassays (review). Biointerphases 12:02D301. https://doi.org/10.1116/1.4978435
7. Gao S, Guisan JM, Rocha-Martin J (2022) Oriented immobilization of antibodies onto sensing platforms – a critical review. Anal Chim Acta 1189:1–24. https://doi.org/10.1016/j.aca.2021.338907
8. Vashist SK, Schneider EM, Lam E, Hrapovic S, Luong JHT (2014) One-step antibody immobilization-based rapid and highly sensitive sandwich ELISA procedure for potential in vitro diagnostics. Sci Rep 4:4407. https://doi.org/10.1038/srep04407
9. Bergeron AB, Bortz CJ, Rossi A (2018) Corning® medium and high binding ELISA microplates for select target size binding assays. Application note CLS-AN-497, corning incorporated
10. Lu D, Yang C, Liu Z (2012) How hydrophobicity and the glycosylation site of Glycans affect protein folding and stability: a molecular dynamics simulation. J Phys Chem B 2012(116):390–400. https://doi.org/10.1021/jp203926r
11. Subedi GP, Barb AW (2015) The structural role of antibody N-glycosylation in receptor interactions. Structure 23(9):1573–1583. https://doi.org/10.1016/j.str.2015.06.015
12. https://www.agrisera.com/en/info/molecular-weight-and-isoelectric-point-of-various-immunoglobulins.html

13. Choe W, Durgannavar TA, Chung SJ (2016) Fc-binding ligands of immunoglobulin G: an overview of high affinity proteins and peptides. Materials 9:1–17. https://doi.org/10.3390/ma9120994
14. Wang TT (2019) IgG fc glycosylation in human immunity. Curr Top Microbiol Immunol 423:63–75. https://doi.org/10.1007/82_2019_152
15. Jankun J (2011) Protein-based nanotechnology: antibody conjugated with photosensitizer in targeted anticancer photoimmunotherapy. Int J Oncol 39:949–953
16. Bondt A, Rombouts Y, Selman MHJ, Hensbergen PJ et al (2014) Immunoglobulin G (IgG) fab glycosylation analysis using a new mass spectrometric high-throughput profiling method reveals pregnancy-associated changes. Mol Cell Proteomics 13(11):3029–3039
17. Xu H, Lu JR, Williams DE (2006) Effect of surface packing density of Interfacially adsorbed monoclonal antibody on the binding of hormonal antigen human chorionic gonadotrophin. J Phys Chem B 110:1907–1914

Chapter 4

Screening for Antibody Specificity and Sensitivity with ELISA

Michael J. Halat and Emily Gravens

Abstract

Monoclonal antibodies (MAbs) can be used to detect and quantify protein biomarker antigens (Ag). Systematic screening with an enzyme-linked immunosorbent assay (Butler, J Immunoass, 21(2–3): 165–209, 2000) [1] can be used to identify matched Ab–Ag pairs. A method is described for identifying MAbs that recognize cardiac biomarker creatine kinase isoform MB. Cross-reactivity with skeletal muscle biomarker creatine kinase isoform MM and brain biomarker creatine kinase isoform BB is also examined.

Key words Antibody, Antigen, Sensitivity, Specificity, Cross–reactivity

1 Introduction

The use of monoclonal antibodies (MAbs) in research, diagnostic, and therapeutic applications has expanded since Kohler and Milstein introduced hybridoma technology in 1975 [2, 3]. The selection of a MAb that reacts with the antigen (Ag) of interest (specificity) to establish a lower limit of detection (sensitivity) can be a daunting task if not approached in a systematic manner. The use of passive binding of an Ag to a microplate well surface can simplify the screening process since the law of mass action governs the interaction of antibody (Ab) and Ag, where K is the equilibrium constant [4, 5]:

$$\text{Ab} + \text{Ag} \rightleftharpoons \text{AbAg}$$

$$K = \frac{[\text{AbAg}]}{[\text{Ab}][\text{Ag}]}$$

In this method, a known mass of Ag is adsorbed to a microplate well via hydrophobic binding (Fig. 1a). Open sites on the plastic are blocked with bovine serum albumin (BSA) to prevent background interference from nonspecific binding to the surface of the

Robert S. Matson (ed.), *ELISA: Methods and Protocols*, Methods in Molecular Biology, vol. 2612,
https://doi.org/10.1007/978-1-0716-2903-1_4,

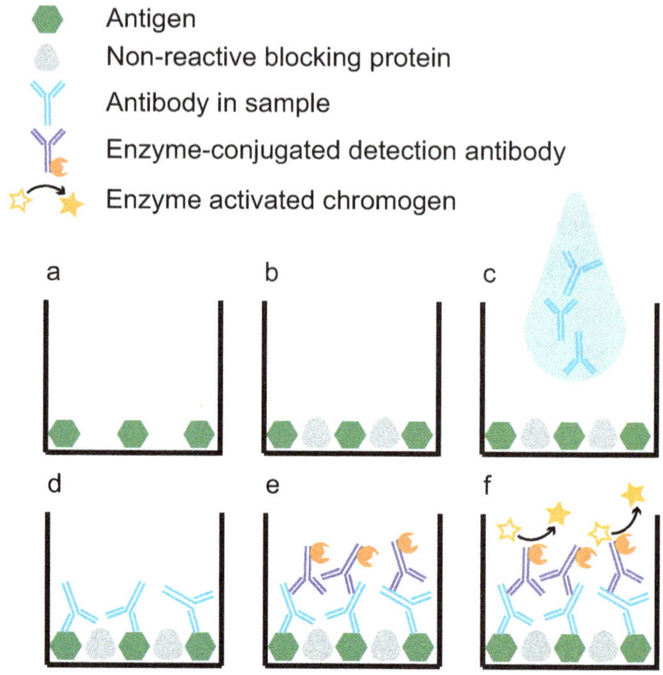

Fig. 1 Illustration of antibody capture by ELISA. Panel (**a**) shows the antigen passively coated to the solid phase. Panel (**b**) shows a nonreactive protein (such as BSA) adsorbed to the open sites on the solid phase. Panel (**c**) shows the addition of a sample containing antibodies. Panel (**d**) shows the specific antibodies bound to the antigen; nonspecific antibodies are washed away. Panel (**e**) shows the enzyme-conjugated detection antibody binding to the primary antibody. Panel (**f**) shows the activation of the chromogen

microplate (Fig. 1b). Serial dilutions of MAb are added to the microplate wells (Fig. 1c). The Ab will bind the Ag if the Ab is specific for the Ag. An Ab that recognizes one Ag is known as monospecific. Unbound Ab will be washed out, leaving only bound Ab in the well (Fig. 1d). A second Ab, conjugated to an enzyme, is used for detection. The second Ab is specific for the species of the Ag-bound Ab. When the Ab -enzyme conjugate is added to the well, it will bind to the primary Ab if present; excess conjugate is washed out (Fig. 1e). A chromogenic substrate is added to the well and is activated by the enzyme. The activated chromogen can be measured as optical density (OD) (Fig. 1f). OD is positively correlated to the amount of MAb bound to the Ag. This technique allows for quick screening of MAbs for desired specificity or undesirable cross-reactivity, as well as comparison of relative sensitivity between two or more MAbs. The following method describes the screening of a single-mouse antihuman CK-MB MAb for specificity to CK-MB Ag and cross-reactivity to CK-MM and CK-BB Ags.

2 Materials

For convenience, prepared and ready-to-use ELISA buffers and diluents were obtained from ImmunoChemistry Technologies, LLC [6]; descriptions of alternative solutions are included in Subheading 4.

1. Monoclonal antibodies for screening: Mouse monoclonal anti-human CK-MB Ab (Omega Biologicals, Inc., Bozeman, MT, USA). Store at 2–10 °C.
2. Antigens of interest: Human CK-BB, human CK-MB, human CK-MM (Omega Biologicals, Inc.). Store at 2–10 °C.
3. Water: reverse osmosis (RO) or distilled.
4. Antigen coating buffer: 5× solution. Store at room temperature. Prepare 1× dilution in water as needed pH (1×) 8.1–8.6 (*see* **Note 1**).
5. General block: A ready-to-use blocking buffer at pH 7.2–7.6, contains ≤0.1% sodium azide, sodium chloride and phosphates, and BSA. Store at 2–10 °C (*see* **Note 2**).
6. ELISA wash buffer: A 10× buffered protein-free solution. Store at room temperature. Prepare 1× dilution in water as needed (*see* **Note 3**). pH (1×) 7.2–7.6, contains pH stabilizers, salts, and detergents, and ≤1% 2-chloroacetamide.
7. General Serum Diluent: A ready-to-use buffer, pH 7.2–7.6. Contains BSA and antimicrobial agents. Store at 2–10 °C (*see* **Note 4**).
8. Goat antimouse IgG Fc horseradish peroxidase (HRP) conjugate: 1 mg/mL in Tris buffer, pH 4.2. Store at 2–10 °C (*see* **Note 5**).
9. Antigen-down HRP conjugate stabilizer: A 5× buffered solution, pH 7.2–7. Store at 2–10 °C; prepare 1× dilution in water as needed (*see* **Note 6**).
10. Tetramethylbenzidine (TMB) 1-component HRP microwell substrate: A ready-to-use solution, pH 3.35–3.75. Store at 2–10 °C (*see* **Note 7**).
11. Stop solution for TMB substrates: A ready-to-use solution containing 3–7% maleic acid, pH ≤ 2.0. Store at room temperature (*see* **Note 8**).
12. 96-Well ELISA microplates.
13. Humidity chamber (*see* **Note 9**).
14. Multichannel pipette, 8-channel or 12-channel capable of pipetting 100 μL.
15. Pipettes capable of pipetting 10–100 μL, 100–1000 μL, and 1–10 mL.

16. Disposable pipette tips.
17. Disposable tubes and vials for volumes of 1–5 mL, 5–15 mL, and 15–50 mL.
18. Disposable microplate seals or sealing film.
19. Reagent reservoirs (Fisher Scientific, Cat# 21-380-3).
20. Multichannel reagent reservoirs (Fisher Scientific, Cat# NC1670686).
21. Microplate reader capable of OD measurements of 96-well plates at 450 nm.
22. Microplate washer or squirt bottle.

3 Methods

Carry out all procedures at room temperature unless otherwise specified.

3.1 Determine Test Conditions and Prepare Ag-Coated Plate

1. Design a 96-well plate testing template. Ensure space for blanks and samples in duplicate. Table 1 shows a testing design allowing for six conditions to be tested on a single 96-well plate with a six-point serial dilution of Ab (*see* **Note 10**).
2. Bring Ag coating buffer 5× and Ags to room temperature (*see* **Note 11**). Ensure the Ag coating buffer is fully dissolved and free of crystals by gentle swirling.
3. Prepare 25 mL of 1× Ag coating buffer by diluting the 5× Ag coating buffer with water (*see* **Note 12**). Combine 5 mL of the 5× Ag coating buffer and 20 mL of water.
4. Prepare Ag dilutions in 1× Ag coating buffer. Prepare 3 mL each of CK-MB, CK-BB, and CK-MM Ags at 2 μg/mL and 6 μg/mL in 1× Ag coating buffer (*see* **Note 13**).
5. Transfer the prepared Ag dilutions and approximately 2.5 mL of blank 1× Ag coating buffer to separate and clean reagent reservoirs. With a multichannel pipette, add 100 μL of solution per well according to the test template (*see* Table 1 and **Note 14**). Cover the microplate with an adhesive plate seal.
6. Place the coated and sealed microplate in a humidity chamber. Allow the Ag-coated plate to incubate overnight at room temperature (*see* **Note 15**).

3.2 Wash and Block Plate

1. Bring general block and 10× ELISA wash buffer to room temperature. Swirl gently to ensure that all solids or crystals are dissolved.
2. Prepare 1 L of 1× ELISA wash buffer by diluting 100 mL of 10× stock with 900 mL of water. Add 30 mL of general block to a clean reagent reservoir.

Table 1
Plate layout for antibody specificity screening by ELISA

<table>
<tr><th></th><th>1</th><th>2</th><th>3</th><th>4</th><th>5</th><th>6</th><th>7</th><th>8</th><th>9</th><th>10</th><th>11</th><th>12</th><th>Ab dilutions</th></tr>
<tr><td>A</td><td colspan="12">No antigen-coated, blank wells</td><td>No ab</td></tr>
<tr><td>B</td><td colspan="2" rowspan="7">Condition 1: coat CK-MB at 2 μg/mL</td><td colspan="2" rowspan="7">Condition 2: coat CK-MB at 6 μg/mL</td><td colspan="2" rowspan="7">Condition 3: coat CK-MM at 2 μg/mL</td><td colspan="2" rowspan="7">Condition 4: coat CK-MM at 6 μg/mL</td><td colspan="2" rowspan="7">Condition 5: coat CK-BB at 2 μg/mL</td><td colspan="2" rowspan="7">Condition 6: coat CK-BB at 6 μg/mL</td><td>No ab</td></tr>
<tr><td>C</td><td>1:1024</td></tr>
<tr><td>D</td><td>1:256</td></tr>
<tr><td>E</td><td>1:64</td></tr>
<tr><td>F</td><td>1:16</td></tr>
<tr><td>G</td><td>1:4</td></tr>
<tr><td>H</td><td>400 ng/mL</td></tr>
</table>

Table shows six antigen coating conditions, a six-point serial dilution of antibody, and blanks with all samples run in duplicate

3. Collect the Ag-coated microplate. Aspirate the well contents, then wash the wells by adding 300–400 μL of 1× ELISA wash buffer to each well using a microplate washer or squirt bottle. Wait 60 s with the wash buffer in each well, then aspirate contents. Repeat the wash, then aspirate the well contents. The remaining 1× ELISA wash buffer can be stored at room temperature for use the following day. Do not allow the wells to dry out; have the next reagent prepared for loading when the wash is complete.
4. With a multichannel pipette, add 300 μL of general block to each well. Cover the microplate with an adhesive plate seal.
5. Place the blocked and sealed microplate in a humidity chamber and allow the plate to incubate overnight at room temperature (*see* **Note 16**).

3.3 Incubate with Ab

1. Bring the General Serum Diluent and Ab solution to room temperature.
2. Dilute the Ab solution to the highest testing concentration with General Serum Diluent (*see* **Note 17**). For the design described in Table 1, prepare 3.6 mL of 400 ng/mL Ab (if using a standard reagent reservoir) or 2.2 mL of 400 ng/mL Ab (if using a multichannel reagent reservoir) (*see* **Note 18**).
3. Prepare fourfold serial dilutions of the Ab in General Serum Diluent (Table 1). If using standard reagent reservoirs: add 2.7 mL of General Serum Diluent to five clean, separate, labeled vials, then add 0.9 mL of 400 ng/mL Ab solution to the first vial and mix well. Add 0.9 mL from the first vial to the second vial, mix well, and so on for the remaining three vials. If using a multichannel reagent reservoir: add 1.65 mL of General Serum Diluent to slots 2–6, then add 2.2 mL of 400 ng/mL Ab solution to slot 1. Add 550 μL of 400 ng/mL Ab solution to slot 2 and mix well. Add 550 μL from slot 2 to slot 3, mix well, and so on for the remaining three slots.
4. Add 3.6 mL of General Serum Diluent to a clean, separate reagent reservoir.
5. Collect the coated and blocked microplate. Empty the wells by either aspirating out the blocking solution or gently turning the plate upside down over a sink or basin and dispelling the well contents with a firm flick of the wrist. Blot the empty plate over paper towels.
6. With a multichannel pipette, add 100 μL per well of the blank General Serum Diluent or Ab dilutions as described in the test template (Table 1). Cover the microplate with an adhesive plate seal.
7. Place the sealed microplate in a humidity chamber and allow the plate to incubate for 75 min at room temperature (*see* **Note 19**).

3.4 Incubate with HRP Conjugate

1. Bring 5× HRP conjugate stabilizer and HRP conjugate to room temperature.
2. Prepare a 1× solution of 5× HRP conjugate stabilizer in water by combining 3 mL of 5× HRP conjugate stabilizer and 12 mL of water.
3. Prepare a 1:100 dilution of HRP conjugate by spiking 10 μL of the HRP conjugate stock into 990 μL of 1× HRP conjugate stabilizer. Prepare a 1:10,000 dilution of HRP conjugate by adding 110 μL of 1:100 HRP conjugate dilution to 10.89 mL of 1× HRP conjugate stabilizer (*see* **Note 20**). Store at room temperature; protect from light until the Ab incubation is complete.
4. After the Ab incubation is complete, wash the plate using either a plate washer or a squirt bottle. Wash the plate three times with 300–400 μL of 1× ELISA wash buffer per well, allowing a one-minute soak between filling the wells and aspirating. After the last wash, aspirate the well contents. Blot the empty plate over paper towels (*see* **Note 21**).
5. Pour the 1:10,000 HRP conjugate dilution prepared in **step 3** into a clean reagent reservoir. With a multichannel pipette, add 100 μL per well of 1:10,000 HRP conjugate dilution (*see* **Note 22**). Cover the microplate with an adhesive plate seal.
6. Place the sealed microplate in a humidity chamber and allow the plate to incubate for 60 min at room temperature.

3.5 Incubate with TMB Substrate

1. Allow TMB substrate to come to room temperature.
2. After the HRP conjugate incubation is complete, wash the plate using either a plate washer or a squirt bottle. Wash the plate four times with 300–400 μL of 1× ELISA wash buffer per well, allowing a one-minute soak between filling the wells and aspirating. After the last wash, aspirate the well contents. Blot the empty plate over paper towels.
3. Add 10.8 mL of TMB substrate to a clean reagent reservoir. With a multichannel pipette, add 100 μL per well of TMB substrate.
4. Allow the plate to incubate for 15 min protected from light.

3.6 Add Stop Solution and Read Plate

1. Power on the plate reader while the TMB is incubating (*see* **Note 23**).
2. Add 10.8 mL of stop solution to a clean reagent reservoir. After a 15-min incubation, collect the TMB-activated plate (*see* **Note 24**). Do not wash the plate or aspirate the well contents. With a multichannel pipette, add 100 μL per well of stop solution.
3. Measure OD of the wells at 450 nm with the plate reader.

Table 2
Optical density data measured at 450 nm from ELISA specificity screening of anti-CK-MB monoclonal antibody (Omega Biologicals, Inc., Cat# 100–31-091)

Anti-CK-MB antibody concentration (ng/mL)	CK-MB 2 μg/mL Average OD 450 nm	CK-MB 6 μg/mL	CK-MM 2 μg/mL	CK-MM 6 μg/mL	CK-BB 2 μg/mL	CK-BB 6 μg/mL
Blank	0.047	0.043	0.043	0.046	0.047	0.043
0	0.045	0.041	0.041	0.047	0.043	0.044
0.390625	0.052	0.057	0.041	0.044	0.045	0.045
1.5625	0.067	0.101	0.045	0.044	0.044	0.043
6.25	0.141	0.271	0.045	0.045	0.044	0.042
25	0.380	0.800	0.044	0.043	0.044	0.043
100	1.088	2.055	0.044	0.045	0.045	0.042
400	2.239	3.664	0.044	0.048	0.043	0.047

Antibody was screened against CK-MB, CK-MM, and CK-BB antigens. Antigens were coated at either 2 μg/mL or 6 μg/mL concentrations (100 μL/well)

3.7 Process and Review Data

1. Process the OD data to compare results of the ELISA screening (*see* Table 2 and **Note 25**).
2. Data can be displayed graphically as OD (y-axis) versus Ab concentration (x-axis). Displaying the x-axis in log-scale may allow easier visualization of the linear range (Fig. 2).
3. The data reported in Table 2 and Fig. 2 clearly demonstrate the anti-CK-MB MAb is selective for CK-MB with no cross-reactivity to CK-MM or CK-BB.
4. The data reported in Table 3 and Fig. 3 were collected with a modified method (*see* **Note 10**). These data demonstrate the comparative sensitivity of six unique MAbs to CK-MB Ag.

4 Notes

1. The composition of the Ag coating buffer can vary, but it should be free of extraneous proteins or detergents, which may disrupt the adsorption of the Ag to the microplate well [7]. Possible compositions include 10 mM phosphate-buffered saline (PBS) at pH 7.2; 20 mM Tris–HCl at pH 8.5; 50 mM carbonate at pH 9.6. Protein coating is more efficient when the buffer pH is 1–2 pH units higher than the pI of the protein [8].
2. A blocking buffer contains a nonreactive protein that will bind to any spaces on the microwell surface that remain available after the Ag coating; this will prevent the nonspecific binding of antibodies or the detection reagent in later steps. Possible

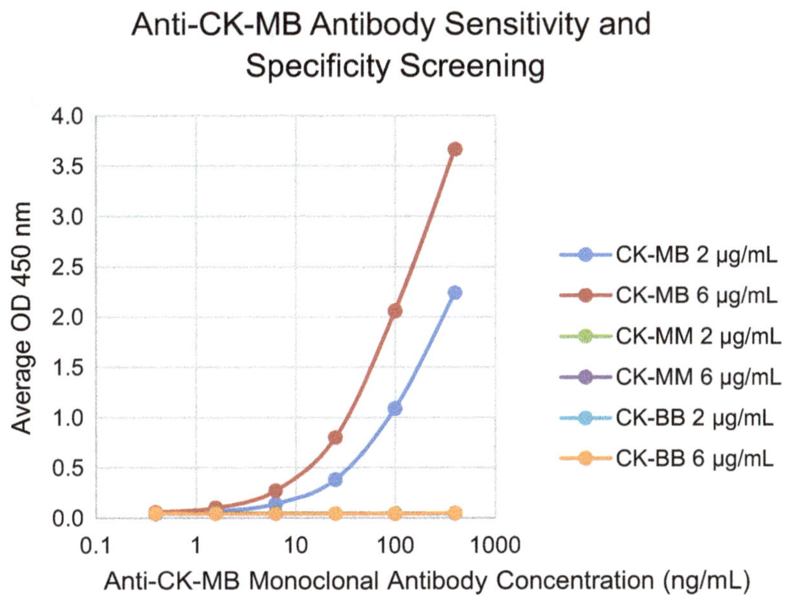

Fig. 2 Graphical depiction of optical density data from antibody specificity screening by ELISA. The antibody tested (Omega Biologicals, Inc., Cat# 100-31-091) showed specificity for CK-MB antigen with no cross-reactivity to CK-MM or CK-BB antigens. Raw data are reported in Table 2

Table 3
Optical density data measured at 450 nm from ELISA sensitivity screening of six monoclonal antibodies (Omega Biologicals, Inc., antibodies are listed by catalog number)

Antibody concentration (ng/mL)	CK-MB antigen coated at 6 μg/mL					
	Cat# 100-31-061 Average OD 450 nm	Cat# 100-31-091	Cat# 100-31-121	Cat# 100-31-151	Cat# 100-31-181	Cat# 100-31-211
Blank	0.045	0.043	0.041	0.053	0.045	0.046
0	0.043	0.041	0.042	0.044	0.042	0.043
0.390625	0.059	0.057	0.057	0.044	0.048	0.043
1.5625	0.107	0.101	0.100	0.046	0.064	0.043
6.25	0.289	0.271	0.280	0.053	0.105	0.044
25	0.832	0.800	0.794	0.079	0.284	0.048
100	2.040	2.055	1.652	0.176	0.586	0.061
400	3.208	3.664	2.392	0.420	0.855	0.093

Antibodies were screened against CK-MB antigen. The antigen was coated at 6 μg/mL (100 μL/well)

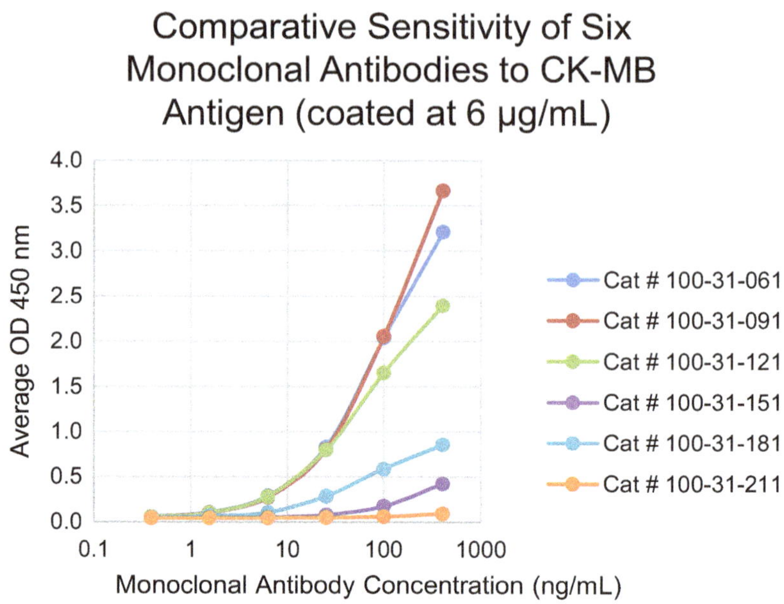

Fig. 3 Graphical depiction of optical density data from antibody sensitivity screening by ELISA. The six antibodies tested show varying degrees of sensitivity to CK-MB antigen. Raw data are reported in Table 3

compositions include 3% BSA in PBS; 5 wt%/vol nonfat dry milk in PBS; 5 wt%/vol nonfat dry milk and 0.2% Tween 20 in PBS; 0.2% Tween 20 in PBS. The addition of 0.02% sodium azide will inhibit bacterial growth during storage. If using alkaline phosphatase (ALP) as the reporter, prepare in a Tris buffer instead of PBS [8].

3. A wash buffer is used to rinse the microplate wells in between steps to remove unbound reagents. The wash buffer can be a PBS buffer or 150 mM NaCl with 0.05% Tween 20 [8]. Sufficient washing will remove interferents and reduce nonspecific binding and high background signals. The number of washes, volume of wash, and soak time can be increased as needed. HRP is inactivated by sodium azide. Efficient washing is critical to assay performance.
4. Samples can be diluted in a PBS buffer with 3% BSA [7].
5. The Ab enzyme-conjugate should be chosen for specificity to the primary Ab being screened. For example, choose a goat antimouse IgG conjugate if the sample contains mouse MAbs or a goat antihuman IgG if the sample contains human antibodies. This method describes the use of HRP as the enzyme; this is the most commonly used enzyme in ELISA formats and many HRP conjugates are commercially available. HRP is commonly used with TMB as the chromogen in ELISA. 2,2′-azino-bis(3-ethylbenzothiazoline-6-sulfonic acid)

(ABTS) and *o*-phenylene diamine (OPD) are also used as chromogens. Of these, TMB is the most sensitive. ALP is another popular enzyme for ELISA formats. ALP is commonly used with *p*-nitrophenyl phosphate (PNPP) as the chromogen [7, 8].

6. A general enzyme conjugate dilution buffer is 3% BSA in a phosphate or citrate buffer with or without 0.05% Tween. HRP is more stable in a citrate buffer than a phosphate buffer and has a broad pH range of optimal enzymatic activity, pH 4.0–8.0. Avoid using sodium azide as it can inhibit HRP activity. ALP has optimal enzymatic activity between pH 8–10. Because ALP requires Mg^{2+} for enzymatic activity, EDTA ethylenediaminetetraacetic acid (EDTA) can inhibit activity. ALP is not affected by sodium azide. Remember that PBS interferes with ALP [8].
7. Ready-to-use TMB substrate solutions are available from several suppliers. In the presence of HRP, TMB is converted to a blue color; the color intensity can be measured at 650 nm. If ABTS is used instead, HRP will convert ABTS to a green color with maximum absorbance at 410 nm and 650 nm [7]. HRP will convert OPD to a brown color with maximum absorbance at 450 nm. ALP will convert PNPP to a yellow color with maximum absorbance at 405 nm. These other chromogens can be found as ready-to-use solutions or tablets from suppliers.
8. A stop solution is used to halt the enzymatic reaction. A 1 M solution of sulfuric acid, H_2SO_4, will halt the TMB conversion and change that color from blue to yellow; the color intensity can be measured at 450 nm. An ALP-PNPP reaction can be stopped with 0.1 M solution of EDTA; the color will still be yellow [7]. The chromogen supplier may recommend an alternate stop solution; follow the supplier's recommendations.
9. The humidity chamber can be any closed container lined with a damp paper towel; rest the microplate on a piece of glass or plastic so that it does not rest directly on top of the damp paper towel. Incubation in a humid atmosphere helps to reduce evaporation from the wells. Evaporation can be uneven across the surface of the plate and can negatively impact reproducibility.
10. The method described screens one Ab against three Ags at two Ag coating concentrations; this method could easily be modified to screen any six conditions. For example, one Ab could be screened against six Ags, or six antibodies against one Ag. Further modification to screen 12, 18, or 36 conditions could be achieved by reducing the six-point serial dilution of Ab to 3, 2, or 1 Ab concentrations, respectively. The six-point

dilution is useful for comparative sensitivity screening and determining the optimal reporting range.

11. Displacement pipettes are most accurate when pipetting solutions that are at room temperature. Pipetting cold solutions can affect the accuracy of the results.
12. Prepare as much Ag solution as needed for the current experiment. To calculate the minimum volume needed, assume 100 μL of Ag coating buffer per well, plus 1.2–1.5 mL overage per Ag coated; this overage will ensure adequate volume in the reagent reservoir.
13. Antigen coating concentrations can range from 1 to 10 μg/mL with coating volumes of 50–100 μL/well [8]. The method described uses 2 μg/mL and 6 μg/mL Ag coating concentrations with a coating volume of 100 μL/well for screening purposes.
14. To prevent contamination, always use clean pipette tips for each dilution or buffer. Some users will use the same pipette tips if going from a lower concentration to a higher concentration, but reproducibility is best when pipette tips are changed for each solution.
15. Antigen coating incubation temperatures and times can vary widely, from overnight at 4 °C, to 1 h at 37 °C. Typically, a longer incubation time leads to increased sensitivity [7, 8]. Some Ags are temperature sensitive [8]. If higher throughput is desired and sensitivity is not a concern, Ag coating incubation could be reduced to 2–4 h at room temperature.
16. Blocking is a critical step for avoiding background interference. Incubation times can vary from 2 h to overnight.
17. The Ab concentration required to produce a signal in the linear range of the assay can vary. A good starting point for screening is 5–500 ng/mL of Ab. If using serum, ascites, or cell culture supernatant instead of purified Ab, the Ab concentration may be estimated.
18. The use of a multichannel reagent reservoir can make it easier and faster to load the ELISA plate, and it can reduce the total volume of sample needed. Determine the overage needed for the reagent basin in use. For this method, 1.2–1.5 mL overage is required if using a standard reagent reservoir and a 0.4 mL overage is required if using a multichannel reagent reservoir. To calculate the minimum volume needed, assume 100 μL of Ab solution per well, plus the overage needed for the reagent reservoir, plus a 1/3 overage to accommodate the fourfold serial dilutions.

19. The Ab incubation time can vary, but it should be at least 1 h at room temperature. Longer incubation times can increase sensitivity but may also increase nonspecific binding. Optimization of Ab incubation time should not be necessary for a screening assay.
20. The concentration of the HRP conjugate can vary; choose a concentration based on the supplier's recommendation. Higher HRP conjugate concentrations can increase sensitivity but can also increase the background signal.
21. Blot the washed plate by rapping the plate firmly against paper towels twice; rotate the plate 180° and repeat. This helps remove any remaining wash buffer and improves reproducibility and precision.
22. Load the plate in the same order each time. For example, if using an 8-channel pipette load from column 1 to column 12 each time; if using a 12-channel pipette load from row A to row H each time. This should be the same order used to wash the plate. This is to avoid having regions of the plate with slightly different incubation times. This is particularly important when incubation times are short.
23. Powering on the plate reader will allow the lamp to warm up for 15 min prior to use.
24. If after 15 min the TMB in the wells with the highest concentration remains a faint blue, continue to incubate for an additional 5 min to try to increase the signal. Longer incubation times can increase the sensitivity but may increase the background signal as well.
25. Rows A and B should exhibit similarly low OD since these rows contain no Ab. Elevated OD in rows A and B indicate high background or nonspecific binding. The cause could be contamination in the solutions and/or instruments, insufficient washing, or insufficient blocking (*see* **Notes 2** and **16**). Instruments should be thoroughly cleaned per the manufacturer's instructions. Fresh solutions should be prepared. The wash method can be modified to reduce nonspecific binding or carryover (*see* **Note 3**). If the OD of row A is low, but the OD of row B is significantly higher, this may be due to the presence of a contaminant in the Ag, leading to cross-reactivity with the enzyme conjugate. Nonspecific binding can be reduced by choosing a high-purity Ag, decreasing the coating concentration of the Ag, or increasing the dilution of the enzyme conjugate. Assay sensitivity may be impacted if the coating concentration is decreased or if the enzyme conjugate solution is further diluted.

References

1. Butler JE (2000) Enzyme-linked immunosorbent assay. J Immunoass 21(2–3):165–209. https://doi.org/10.1080/01971520009349533
2. Kohler G, Milstein C (1975) Continuous cultures of fused cells secreting antibody of predefined specificity. Nature 256:495–497. https://doi.org/10.1038/256495a0
3. Ossipow V, Fischer N (eds) (2014) Monoclonal antibodies, methods and protocols, 2nd edn. Humana Press. https://doi.org/10.1007/978-1-62703-992-5
4. Raghava GPS, Agrewala JN (1994) Method for determining the affinity of monoclonal antibody using non-competitive Elisa: a computer program. J Immunoass Immunochem 15(2): 115–128. https://doi.org/10.1080/15321819408013942
5. van Regenmortel MHV, Azimzadeh A (2000) Determination of antibody affinity. J Immunoass 21(2–3):211–234. https://doi.org/10.1080/01971520009349
6. Immunochemistry Technologies, LLC (2018) Antigen-down ELISA development kit. https://immunochemistry.com/product/antigen-elisa-development-kit/. Accessed 16 Aug 2018
7. Harlow E, Lane D (1988) Immunoassay. In: Antibodies: a laboratory manual. Cold Spring Harbor Laboratory, New York. ISBN: 978-0879693145
8. He J (2013) Practical guide to ELISA development. In: Wild D (ed) The immunoassay handbook, 4th edn. Elsevier Ltd., Oxford. ISBN: 978-0-08-097037-0

Chapter 5

Effective Blocking and Stabilizing Methods Using Synthetic Polymer on ELISA

Tomozumi Noda, Masaru Matsuda, Hirotaka Suzuki, Yusuke Okawa, and Motohiro Mitani

Abstract

In ELISA, blocking reagents and stabilizers are important to improve the sensitivity and/or quantitative nature of the measurement system. Usually, biological substances such as bovine serum albumin and casein are used, but they still have problems such as lot-to-lot differences and biohazard. Here, we describe the methods using a chemically synthesized polymer, BIOLIPIDURE®, as a new blocking agent and stabilizer that can solve these problems.

Key words ELISA, Antibody, Blocking, Stabilizer, Chemically synthesized materials

1 Introduction

For early detection of diseases and rapid determination of infectious diseases, trace components contained in biological samples such as serum, plasma, urine, and saliva must be measured. Furthermore, since biological samples contain many impurities, high sensitivity, accuracy, and precision are required. One such assay widely used for highly quantitative measurement is the enzyme-linked immunosorbent assay (ELISA) using an enzyme-labeled antibody.

However, the sensitivity and/or linearity is reduced due to the nonspecific adsorption of enzyme-labeled antibody, antigen, and biological impurities on the surface of the substrate (e.g., polystyrene 96-well microplate for immunological measurement). Therefore, a blocking reagent is required not only for ELISA but also for Western blot technique, immunostaining, immunochromatography, and flow cytometry [1]. The blocking agent is also used for stabilization of the immobilized antibody. When the immobilized antibody is unstable on the substrate, the reproducibility of ELISA will be decreased.

Robert S. Matson (ed.), *ELISA: Methods and Protocols*, Methods in Molecular Biology, vol. 2612,
https://doi.org/10.1007/978-1-0716-2903-1_5,

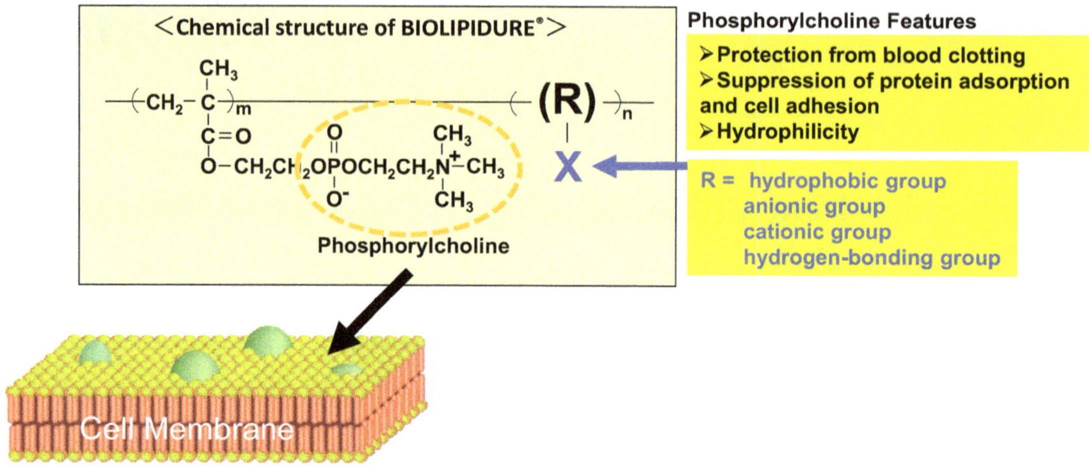

Fig. 1 Chemical structure of BIOLIPIDURE®

In general, bovine serum albumin (BSA), casein, ovalbumin, (OVA), and various animal sera are used as blocking reagents for suppression of nonspecific adsorption and stabilization of immobilized antibody. Additionally, in case the enzyme-labeled antibody is not stable, the reproducibility and reliability of the ELISA will be further reduced [2].

However, blocking reagents or stabilizers derived from animal protein have several problems such as lot-to-lot variation, risk of infection, and denaturation due to repeated freeze-thawing [2]. Therefore, the use of nonprotein blocking reagents is preferable.

BIOLIPIDURE® is an additive for in vitro diagnostics developed by NOF CORPORATION. BIOLIPIDURE® is a water-soluble 2-methacryloyloxyethyl phosphoryl choline (MPC)-based polymer, which is produced by copolymerization of MPC and various co-monomers (*See* **Note 1**). MPC is a compound composed of the methacrylate group and the phosphorylcholine group and has high biocompatibility derived from the phosphorylcholine group, which is the hydrophilic part of phospholipid on the cell membrane of living organisms (*see* Fig. 1). In addition, the methacrylate group enables the synthesis of a great variety of polymers having different physical properties, which is derived from a rich variation of secondary monomers copolymerized with MPC.

NOF CORPORATION has been developing various types of MPC-based polymers that are used in various fields, such as medical devices, eye care, oral care, and skin care, as well as in vitro diagnostics. In general, MPC-based polymers do not exhibit foreign body reaction in humans and suppress interactions with biological objects because their structures sequenced phosphorylcholine group in the side chain are similar to a cell membrane [3]. Therefore, MPC-based polymers also exhibit excellent biocompatibility, including the ability to suppress denaturation of a protein and adsorption of biomolecule, cell, and blood component, which is attributed to the formation of the unique hydration state of the phosphorylcholine group at aqueous interface [4]. Due to these functions, BIOLIPIDURE® is used for applications such as (1) suppression of nonspecific adsorption of protein, (2) stabilization of protein, and (3) sensitization in immunoassays as additives for diagnostics in many different assays, including ELISA. Furthermore, BIOLIPIDURE® is a chemically synthesized material, so it has high stability, no lot-to-lot variation, and no biohazardous risk compared to biological additives such as BSA and casein.

2 Materials

NOF CORPORATION has a variety of BIOLIPIDURE® products depending upon the purpose, target, and assay types, as shown in Table 1. Some of the appropriate types of BIOLIPIDURE® could be chosen in reference to Table 2, and the method to use BIOLIPIDURE® is shown in detail as follows (*see* **Notes 2–4**).

Table 1
BIOLIPIDURE® product list

Product name	Feature	Suggested application
BIOLIPIDURE®-103	Very hydrophilic (neutral)	Latex agglutination Lateral flow test
BIOLIPIDURE®-203	Moderately hydrophobic	Blocking in ELISA/CLEIA Latex agglutination
BIOLIPIDURE®-206	Relatively hydrophobic	Lateral flow test Protein stabilizer
BIOLIPIDURE®-401	Very hydrophilic (anionic)	Lateral flow test Metal surface modification
BIOLIPIDURE®-403	Very hydrophilic (anionic)	
BIOLIPIDURE®-405	Very hydrophilic (anionic)	
BIOLIPIDURE®-802	Relatively hydrophobic	Blocking in ELISA/CLEIA Lateral flow test Protein stabilizer Solubilizer
BIOLIPIDURE®-1002	Moderately hydrophobic	Blocking in ELISA/CLEIA Colorimetric assay Solubilizer
BIOLIPIDURE®-1201	Moderately hydrophobic	Blocking in ELISA/CLEIA Latex agglutination

Table 2
Recommended BIOLIPIDURE® products by application

Assay	Blocking	Stabilizing	Sensitizing
ELISA/CLEIA (microplate)	203, 206, 802, 1002, and 1201	203 and 802	Under evaluation
CLEIA (magnetic beads)	203, 206, and 1002	802	Under evaluation
Lateral flow (with latex)	203, 206, and 802	802	103
Lateral flow (with gold nano-particle)	203, 206, and 802	802	401, 403, and 405

2.1 Blocking Method for Enzyme-Labeled Antibody on Microplate

1. BIOLIPIDURE®-203 blocking solution (0.5 wt%): Prepared from dilution of BIOLIPIDURE®-203 (5 wt%), dilute 1:10 (v/v) with 1× PBS to prepare BIOLIPIDURE®-203 (0.5 wt%).
2. BIOLIPIDURE®-802 blocking solution (0.5 wt%): Prepared from dilution of BIOLIPIDURE®-802 (5 wt%), dilute 1:10 (v/v) with 1× PBS to prepare BIOLIPIDURE®-802 (0.5 wt%).
3. BSA solution: 0.5% in 1× PBS, pH 7.4.
4. 1× PBS: Prepared from dilution of 10× buffer, weigh out 36.3 g $Na_2HPO_4 \cdot 12H_2O$, 2.4 g KH_2PO_4, 80.0 g NaCl, and 2.0 g KCl, dissolve in 850 mL ultrapure water, adjust to pH 7.4 with 6 M HCl, adjust to a final volume of 1000 mL with ultrapure water, and dilute 1:10 (v/v) with water to prepare 1× PBS (*see* **Note 5**).
5. Antibody-HRP conjugate solution: HRP-labeled goat anti-mouse IgG, 1:24,000 dilution in 1× PBS, Bio-Rad, #170-6516.
6. 1× wash buffer: Prepared from dilution of 20× wash buffer, weigh out 5.8 g $Na_2HPO_4 \cdot 12H_2O$, 0.4 g KH_2PO_4, 16.0 g NaCl, and 0.4 g KCl, dissolve in 600 mL ultrapure water, add 10.0 g of polyoxyethylene (20) sorbitan monolaurate to the solution, adjust to a final volume of 1000 mL with ultrapure water, and dilute 1:20 (v/v) with water to prepare (*see* **Note 5**).
7. TMB substrate solution.
8. 1 M sulfuric acid.
9. Microplate: MaxiSorp™ 96-well plate.
10. Pipettor: Single-channel; multichannel.
11. Pipette tips: 1–20 μL; 100–1000 μL; 1–5 mL.
12. Colorimetric plate reader.

2.2 Blocking Method for Enzyme-Labeled Antibody on Magnetic Beads

1. BIOLIPIDURE®-203 blocking solution (1.0 wt%): Prepared from dilution of BIOLIPIDURE®-203 (5 wt%), dilute 1:5 (v/v) with 0.1 M Tris–HCl buffer pH 7.4 to prepare BIOLIPIDURE®-203 (1.0 wt%).
2. BSA solution: 1.0 wt% in 0.1 M Tris–HCl buffer, pH 7.4.
3. 0.1 M Tris–HCl buffer, pH 7.4: prepared from dilution of 1 M Tris–HCl buffer pH 7.4, weigh out 121.1 g of $NH_2C(CH_2OH)_3$(Tris), dissolve in 800 mL ultrapure water, adjust to pH 7.4 with 12 N HCl, adjust to a final volume of 1000 mL with ultrapure water, and dilute 1:10 (v/v) with water to prepare.
4. 0.1 M Tris–HCl buffer, pH 9.5: prepared from dilution of 1 M Tris–HCl buffer pH 9.5, weigh out 121.1 g of $NH_2C(CH_2OH)_3$(Tris), dissolve in 800 mL ultrapure water, adjust to pH 9.5 with 12 N HCl, adjust to a final volume of 1000 mL with ultrapure water, and dilute 1:10 (v/v) with water to prepare.
5. TBS-T: Weigh out 0.05 g polyoxyethylene (20) sorbitan monolaurate and dissolve in 1000 mL of 0.1 M Tris–HCl buffer pH 7.4.
6. Antibody-ALP conjugate: ALP-labeled goat antihuman IgG, 1:10,000 dilution in 0.1 M Tris–HCl buffer pH 7.4, Rockland, #609-1503.
7. CDP-Star solution: CDP-Star, 1:50 dilution in 0.1 M Tris–HCl buffer pH 9.5, Roche, #03715507001.
8. Magnetic beads: Magnosphere™ MX100, JSR Life Sciences.
9. Micro test tube (PP).
10. Magnet.
11. Microplate: Flat-bottom, white, 96-well.
12. Pipettor: Single-channel; multichannel.
13. Pipette tips: 20–200 μL; 100–1000 μL.
14. Chemiluminescent plate reader.

2.3 Sandwich ELISA

1. BIOLIPIDURE®-1002 blocking solution (0.5 wt%): Prepared from dilution of BIOLIPIDURE®-1002 (5 wt%), dilute 1: 10 (v/v) with TBS-T to prepare BIOLIPIDURE®-1002 (0.5 wt%).
2. BSA solution: 0.5 wt% in 0.1 M Tris–HCl buffer pH 7.4.
3. TBS-T: Weigh out 0.05 g polyoxyethylene (20) sorbitan monolaurate and dissolve in 1000 mL of 0.1 M Tris–HCl buffer pH 7.4.
4. 0.1 M Tris–HCl buffer, pH 7.4: Prepared from dilution of 1 M Tris–HCl buffer pH 7.4, weigh out 121.1 g of NH_2C

$(CH_2OH)_3$(Tris), dissolve in 800 mL ultrapure water, adjust to pH 7.4 with 12 N HCl, adjust to a final volume of 1000 mL with ultrapure water, and dilute 1:10 (v/v) with water to prepare.

5. Primary antibody solution: Antihuman IgG antibody, dilute in 0.1 M Tris–HCl buffer pH 7.4 at 2 μg/mL, EXBIO, #11-319-C100.
6. Antigen standard solution: Human IgG, dilute in 0.1 M Tris–HCl buffer pH 7.4 at 10–10,000 ng/mL, Oriental Yeast, #47364000.
7. Secondary antibody solution: ALP-labeled antihuman IgG antibody, 1:10,000 dilution in 0.1 M Tris–HCl buffer, pH 7.4, Rockland, #609-1503.
8. Microplate: MaxiSorp™ 96-well plate.
9. Microplate: Flat-bottom, white, 96-well.
10. Pipettor: Single-channel; multichannel.
11. Pipette tips: 20–200 μL; 100–1000 μL.
12. CDP-Star solution: CDP-Star, 1:50 dilution in 0.1 M Tris–HCl buffer pH 9.5, Roche, #03715507001.
13. Chemiluminescent plate reader.

2.4 Stabilization of Enzyme Conjugate

1. BIOLIPIDURE®-203: 5 wt% aqueous solution.
2. BIOLIPIDURE®-802: 5 wt% aqueous solution.
3. 1× PBS: Prepared from dilution of 20× wash buffer, weigh out 5.8 g $Na_2HPO_4 \cdot 12H_2O$, 0.4 g KH_2PO_4, 16.0 g NaCl, and 0.4 g KCl, dissolve in 600 mL ultrapure water, adjust to a final volume of 1000 mL with ultrapure water, and dilute 1:20 (v/v) with water to prepare (*see* **Note 5**).
4. Screw cap polypropylene microcentrifuge tube.
5. PBS-sucrose solution (10.2 wt%): Weigh out 2.0 g sucrose and dissolve in 17.6 mL of 1× PBS.
6. Antibody-HRP conjugate: HRP-labeled antibody, Bio-Rad, #170-6516.
7. Conical tube: 50 mL.
8. Microplate: Flat-bottom, clear, 96-well.
9. Pipettor: Single-channel; multichannel.
10. Pipette tips: 20–200 μL; 100–1000 μL.
11. TMB substrate solution.
12. 1 M sulfuric acid.
13. Colorimetric plate reader.

2.5 Application of BIOLIPIDURE® for Different Immunoassay-Like Lateral Flow Test

2.5.1 Application for Coronavirus (SARS-CoV-2) IgM/IgG Test Kit

1. Immunochromatography test kit of novel coronavirus (SARS-CoV-2) IgM antibody: RayBiotech Inc., #CG-CoV-IgM-RUO.
2. Immunochromatography test kit of novel coronavirus (SARS-CoV-2) IgG antibody: RayBiotech Inc., #CG-CoV-IgG.
3. COVID-19 serum sample (IgM): IgM-positive, RayBiotech Inc., CoV-PosM-S-100.
4. COVID-19 serum sample (IgG): IgG-positive, RayBiotech Inc., CoV-PosG-S-100.
5. BIOLIPIDURE® solution: Prepared by mixing equal amounts of BIOLIPIDURE®-103 and BIOLIPIDURE®-405.
6. Analyte diluent containing BIOLIPIDURE® solution: Prepared from dilution of BIOLIPIDURE® solution, dilute 1:4 (v/v) with diluent attached to the kit to prepare analyte diluent containing BIOLIPIDURE® solution.

2.5.2 Application for Substances of Coronavirus (SARS-CoV-2) IgG Test Kit

1. Immunochromatography test kit of novel coronavirus (SARS-CoV-2) IgG antibody: RayBiotech Inc., #CG-CoV-IgG.
2. COVID-19 serum sample (IgG): IgG-positive, RayBiotech Inc., CoV-PosG-S-100.
3. Sample solution: Prepared from dilution of COVID-19 serum sample (IgG), dilute 1:7.8 (v/v) with diluent attached to the kit to prepare sample solution.
4. BIOLIPIDURE®-103.

3 Methods

Please carry out all procedures at room temperature unless otherwise specified.

3.1 Blocking Method for Enzyme-Labeled Antibody on Microplate

1. Add 200 μL/well of each blocking solution (BIOLIPIDURE®-203 blocking solution, BIOLIPIDURE®-802 blocking solution, and BSA solution) to the MaxiSorp™ plate.
2. Incubate the plate for 1–2 h.
3. Aspirate and dry in a desiccator for 16–24 h.
4. Add 100 μL/well of antibody-HRP conjugate solution to each well.
5. Incubate the plate for 1 h.
6. Wash the plate five times with 200 μL/well of wash buffer.
7. Add 100 μL/well of TMB substrate solution to each well.

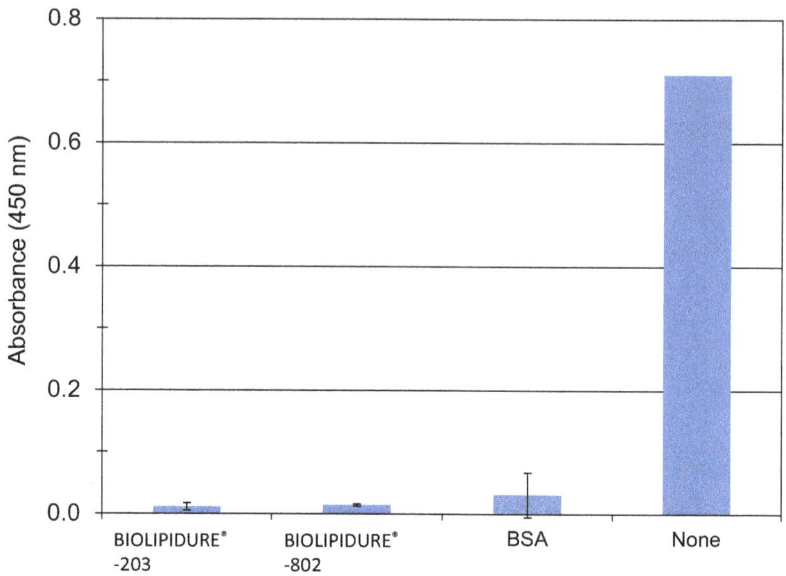

Fig. 2 The result of blocking test of enzyme-labeled antibody on polystyrene plate. BIOLIPIDURE®-203 and BIOLIPIDURE®-802 showed a blocking ability equal to or higher than BSA

8. Incubate the plate for 15 min.
9. Add 50 μL/well of 1 M sulfuric acid.
10. Measure absorbance on a plate reader at 450 nm (Fig. 2).

3.2 Blocking Method for Enzyme-Labeled Antibody on Magnetic Beads

1. Mix equal amounts of 170 μg/mL magnetic beads (Magnosphere™ MX100, JSR Life Sciences) and BIOLIPIDURE®-203 blocking solution or BSA solution.
2. Incubate the magnetic beads for 1 h.
3. Collect the beads using magnet.
4. Aspirate the supernatant and wash with 200–500 μL of TBS-T.
5. Repeat **steps 4** and **5** a total of three times.
6. Add 100 μL/well of antibody-ALP conjugate solution.
7. Incubate the magnetic beads for 1 h.
8. Wash the magnetic beads three times with TBS-T (*see* **steps 4** and **5**).
9. Add 100 μL/well of CDP-Star solution.
10. Incubate the plate for 10–15 min.
11. Transfer the solution to new well of white 96-well plate (flat bottom).
12. Measure relative light unit on plate reader (Fig. 3).

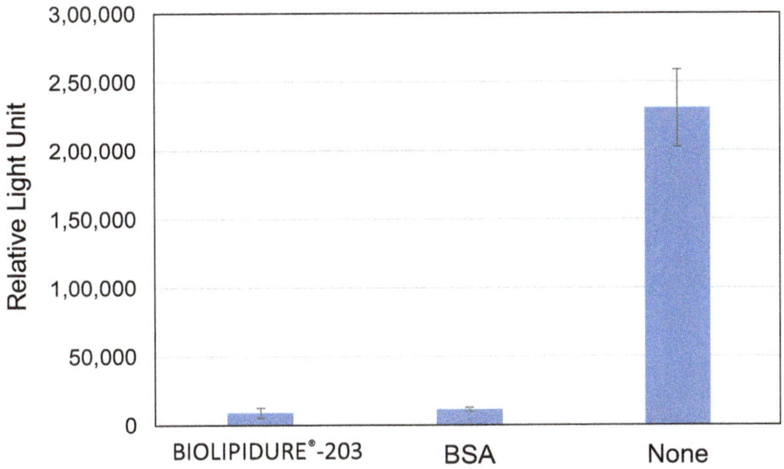

Fig. 3 The result of blocking test for enzyme-labeled antibody on magnetic beads. On magnetic beads, BIOLIPIDURE®-203 showed a blocking ability equal to or higher than BSA

3.3 Sandwich ELISA

1. Add 100 μL/well of primary antibody solution to the MaxiSorp™ plate.
2. Incubate the plate for 1 h.
3. Wash the plate three times with 0.1 M Tris–HCl buffer pH 7.4.
4. Add 200 μL/well of each blocking solution (BIOLIPIDURE®-1002 blocking solution and BSA solution) to the plate.
5. Incubate for 1 h.
6. Aspirate all blocking reagents.
7. Add 100 μL/well of antigen standard solution to the plate.
8. Incubate for 1 h.
9. Wash the plate three times with TBS-T.
10. Add 100 μL/well of secondary antibody solution to the plate.
11. Incubate the plate for 1 h.
12. Wash the plate three times with TBS-T.
13. Add 100 μL/well of CDP-Star solution to the plate.
14. Incubate the plate for 10–15 min.
15. Transfer the solution to new well of white 96-well plate (flat bottom).
16. Measure relative light unit on plate reader (Fig. 4).

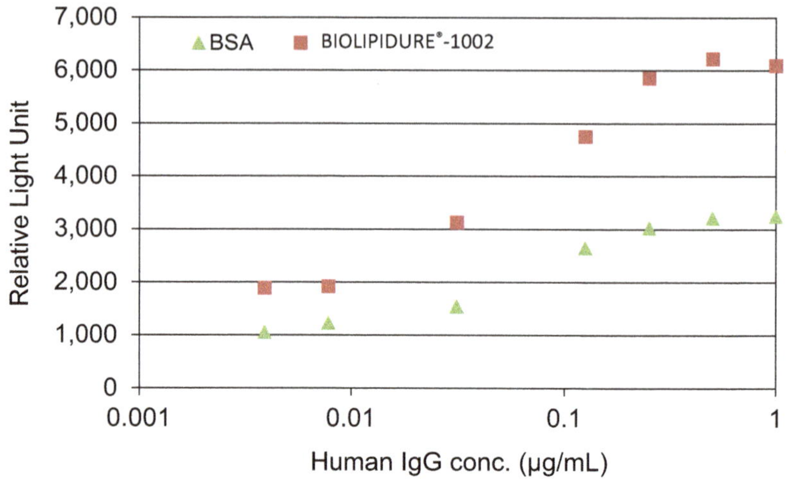

Fig. 4 The result of human IgG measurement system. BIOLIPIDURE®-1002 enhanced the signal intensity in sandwich ELISA

3.4 Stabilization of Enzyme Conjugate

1. Add 19.6 mL of PBS-sucrose solution to conical tube.
2. Add 0.4 mL of BIOLIPIDURE®-203 or BIOLIPIDURE®-802 solution and 1 μL antibody-HRP conjugate to the tube.
3. Dispense 2 mL of the solution into 2 mL screw cap polypropylene microcentrifuge tubes.
4. Store at 4 °C for more than 12 months.
5. Dispense 8 μL of the solution into a well of 96-well plate (flat bottom, clear).
6. Add 100 μL/well of TMB substrate solution to each well.
7. Incubate the plate for 10 min.
8. Add 50 μL/well of 1 M sulfuric acid to stop the enzymatic reaction.
9. Measure absorbance on a plate reader at 450 nm (Fig. 5).

3.5 Application of BIOLIPIDURE® for Different Immunoassay-Like Lateral Flow Test

3.5.1 Application for Coronavirus (SARS-CoV-2) IgM/IgG Test Kit

1. Prepare 1:9.8 (v/v) sample solution of COVID-19 serum sample (IgM or IgG) and analyte diluent containing BIOLIPIDURE® solution in a suitable vessel.
2. Add 2–3 drops of the sample solution to sample dropping point.
3. Incubate for 15 min.
4. Check the result visually (Fig. 6).

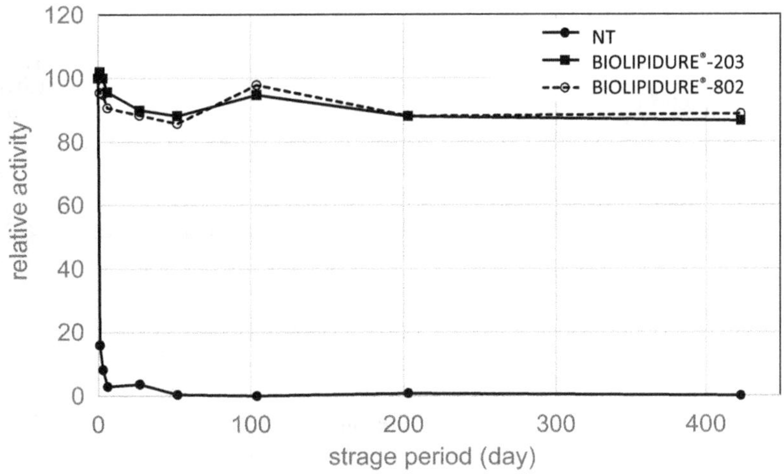

Fig. 5 The result of stabilization of enzyme conjugate. BIOLIPIDURE®-203 or BIOLIPIDURE®-802 kept the activity more than 400 days at 4 °C. In the case of NT (containing only phosphate and sucrose), the activity went down immediately

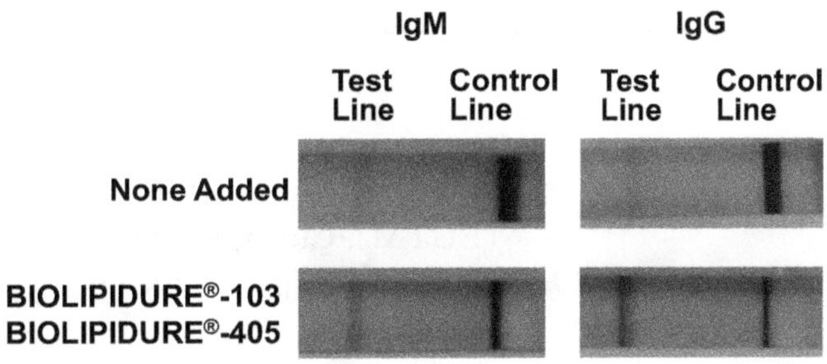

Fig. 6 The result of immunochromatography for the rapid detection of the novel coronavirus (SARS-CoV-2) IgM/IgG antibody. The test line with the addition of BIOLIPIDURE®-103 and BIOLIPIDURE®-405 turned red more clearly. Therefore, the sensitivity of immunochromatography test kit for the rapid detection of the novel coronavirus (SARS-CoV-2) IgM/IgG antibody was significantly improved by adding BIOLIPIDURE®-103 and BIOLIPIDURE®-405

3.5.2 Application for Substances of Coronavirus (SARS-CoV-2) IgG Test Kit

1. Disassemble the detection cassette of the kit and remove sample and conjugate pad.
2. Treat the sample/conjugate pad with 100 μL of BIOLIPIDURE®-103 for the sample pad and 10 μL of BIOLIPIDURE®-103 for the conjugate pad.
3. After drying, reassemble the kit.
4. Add 2–3 drops of the sample solution to the sample dropping point.
5. Incubate for 10 min.
6. Check the result visually (Fig. 7).

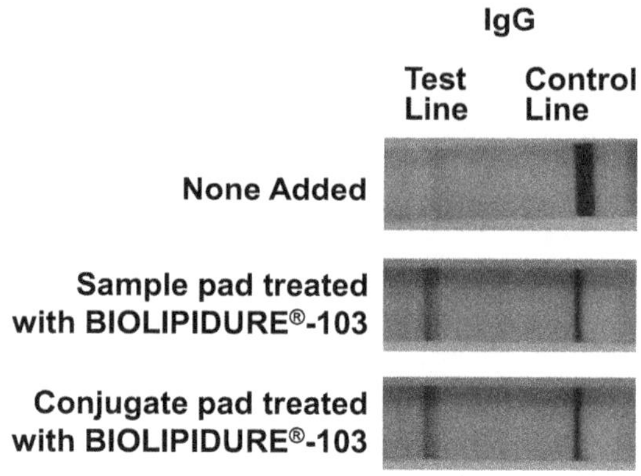

Fig. 7 The result of immunochromatography for the rapid detection of the novel coronavirus (SARS-CoV-2) IgG antibody. The test line treated the sample/conjugate pad with BIOLIPIDURE®-103 in advance turned red more clearly. Therefore, the sensitivity of immunochromatography test kit for the rapid detection of the novel coronavirus (SARS-CoV-2) IgG antibody was significantly improved by treating the sample/conjugate pad with BIOLIPIDURE®-103 in advance

4 Notes

1. BIOLIPIDURE® is a registered trademark of NOF CORPORATION in Japan, the United States, the European Union (EUTM), China, South Korea, and other registered countries.
2. BIOLIPIDURE® with 10–50 times dilution is recommended for use. For initial evaluation, 10 times dilution is recommended. Note that BIOLIPIDURE® is 5 wt% aqueous solution; therefore, the final concentration of BIOLIPIDURE® should be 0.1–0.5 wt% in case of 10–50 times dilution.
3. BIOLIPIDURE® contains no preservative; therefore, use a clean bench when opening the container and store it in the refrigerator.
4. BIOLIPIDURE® is stable chemically but NOF's guarantee period is 1 year after shipment at 2–10 °C with unopened condition.
5. 20× PBS, 1× PBS, 20× wash buffer, and wash buffer do not contain antibiotics. Therefore, be careful of microbial contamination. For storage, sterilize with filtration (pore size is 0.22 μm).

References

1. Sakaki S, Iwasaki Y, Nakabayashi N, Ishihara K (2000) Water-soluble 2-methacryloyloxyethyl phosphorylcholine copolymer as a novel synthetic blocking reagent in immunoassay system. Polymer J 32(8):637–641
2. Sakaki S, Nakabayashi N, Ishihara K (1999) Stabilization of an antibody conjugated with enzyme by 2-methacryloyloxyethyl phosphorylcholine copolymer in enzyme-linked immunosorbent assay. J Biomed Mater 47(4):523–528
3. Ishihara K (2010) MMJ the Mainichi medical Journal, "Medical Forest: Prospects for the Future (2): New Material for Medical Use "MPC Polymer"" 6:68–70
4. Ishihara K, Mu M, Konno T et al (2017) The unique hydration state of poly (2-methacryloyloxyehtyl Phosphorylcholine). J Biomater Sci Polym Ed 28:884–899

Chapter 6

Gas Plasma Surface Modification for Biological Assays

Khoren Sahagian

Abstract

Surface chemistry plays an important role in the adsorption and immobilization of enzymes and antibodies. Gas plasma technology performs surface preparation that assists in the attachment of molecules. Surface chemistry helps to manage a material's wetting, joining, or the reproducibility of surface interactions. There are numerous examples of commerically available products that utilize gas plasma in their manufacturing process. Examples of products treated by gas plasma are well plates, microfluidic devices, membranes, fluid dispensers, and some medical devices. This chapter presents an overview of gas plasma technology and provides a guide for using gas plasma for designing surfaces in product development or research.

Key words Plasma, Surface treatment, Functionalization, Immobilization, Conjugation, Hydrophobic, Hydrophilic, Gas plasma, Vacuum plasma, Atmospheric plasma, Amine, Carboxyl, Hydroxyl

1 Introduction

Gas plasma processes utilize reactive gas streams for the purpose of activating surface chemistry. Most notably, active surface chemistry can help to attach molecules like enzymes, antibodies, proteins, gels, ligands, and receptors. On the other hand, plasma may also be used to inactivate a surface. Thus, a surface may repel or resist binding. The final state of the surface may be determined by the type of plasma chemistry being used. A wide range of material types may be treated by plasma including metals, ceramics, and polymers. Other desirable attributes of plasma processing are the reproducibility, low operating cost and low waste generation.

Surface engineering plays a growing role in the manufacture of diagnostic and microfluidic devices. Plasma processes may be implemented as a batch or in-line operation. Such tools may be adapted for the treatment of a variety of material form factors such as injection moldings, extruded profiles, or nonwoven (*see* Figs. 1 and 2). Surface chemistry helps to manage the material's wetting, joining, and the reproducibility of surface interactions. There can

Robert S. Matson (ed.), *ELISA: Methods and Protocols*, Methods in Molecular Biology, vol. 2612,
https://doi.org/10.1007/978-1-0716-2903-1_6,

Fig. 1 Examples of atmospheric and low-pressure plasma systems manufactured by Plasmatreat

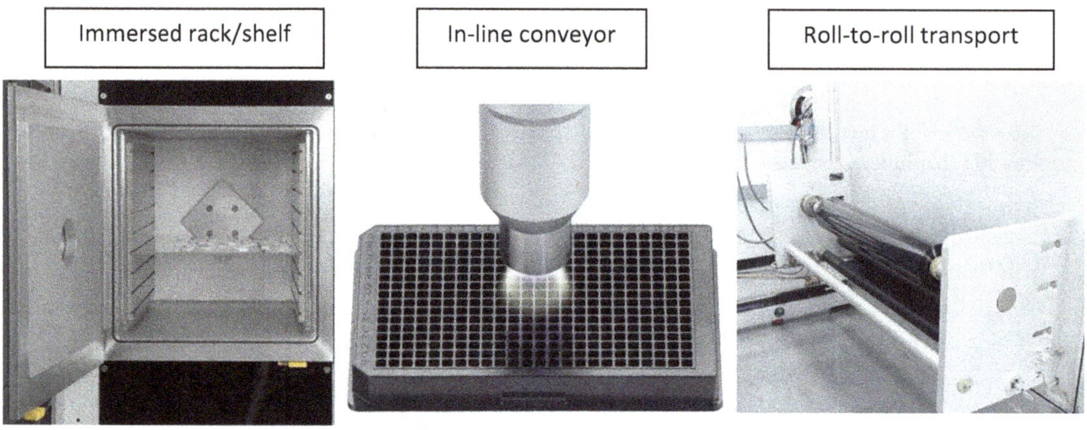

Fig. 2 System configurations shown from left to right: Plasmatreat Aurora 350, Plasmatreat RD1004, and Plasma Science roll-to-roll system

be two directions of interaction between the plasma and a surface: additive and subtractive. Most types of plasma technologies perform a subtractive operations like etching. Surface etching might also be referred to as plasma cleaning. The destructive intent of an etching plasma makes it less unsuitable for additive processes like deposition. Etching equipment might also be too aggressive to be compatible with some thermally sensitive materials such as polymers. The plasma energy and plasma energy source are important considerations for determining the final application requirements.

There are a few ways to generate charged particles and disassociated chemical compounds in a plasma state. These moieties may be ionized using high-voltage arc, strong heat, or electromagnetic field. Adequate energy must be applied to a plasma source; otherwise, the plasma may become unstable. The precise method of creating a plasma will depend on the intended surface chemistry and manufacturing process; i.e., in-line or batch treatment. A plasma generated under atmospheric pressure has a distinct in-line advantage for achieving selective area treatment. A batch system, on the other hand, immerses the objects wholly within a plasma. The chamber has a distinct advantage for processing complex or porous geometries. Batch and in-line processing both allow for precise production controls in the engineering of highly tailored surfaces.

Plasma treatment modifies the surfaces without affecting the bulk material properties. The method provides an inexpensive means of customizing surface chemistry without a heavy reliance on wet chemicals or the need for specialty materials. In most situations, a low-cost or commodity material may be re-engineered into high-performance surface applicable in cell culture, selective filtration, or molecular discovery. The author provides guidance herein for selection of equipment, chemistry, and process variables relevant to the development of surface chemistry for biological assays.

2 Materials

Table 1 provides a list of materials that are commercially plasma treated for biological assays. An initial criterion for material selection should often consider the bulk behavior (*see* **Note 1**). For example, optical transmission, processability, biocompatibility, and gas or vapor transmission. The ideal bulk property and surface property may sometimes be at odds with one another. Fortunately, plasma processes enable downstream redesign of surface characteristics. Nonpolar materials make it difficult to predictably wet reagents and liquids. Poor wetting and surface heterogeneity can account for measurable inconsistency or error. The surface may undergo molecular re-engineering to achieve desired wetting envelope or create chemical linkers for covalent bonding. The intensity of the effect may be tailored to maximize a desired molecular interaction.

There are multiple hardware components considered common to systems that produce plasma. The equipment used for the processes described in this chapter was manufactured by Plasmatreat. The following is a list of common components in equipment used by the authors [1].

1. Ionization chamber: The portion of the system that generates the plasma. The chamber volume may be either sealed or open to atmosphere (*see* **Note 8**). When parts are loaded inside the vessel, this is referred to as primary plasma treatment. Otherwise, the parts may be placed outside the ionization volume

Table 1
Materials used in biological assay development

Material	Product form	Plasma process
Cyclo-olefin polymer or copolymer	Microtiter plate	Stable wetting, functional surfaces
Polystyrene	Microtiter plate Pipettes	Stable wetting, functional surfaces
Polycarbonates	Microtiter plate	Stable wetting, functional surfaces
Glass (borosilicate, quartz)	Pipettes	Functional surfaces
Fluoropolymers	Dispensing components Transfer membranes	Stable wetting
Polypropylene	Pipettes	Stable wetting
PEEK	Dispensing components	
Sintered polyethylene	Wicks, capillaries, filtration Transfer membranes	Hydrophilicity, hydrophobicity Enhance binding Enhance flow
Nitrocellulose	Wicks, capillaries, filtration Transfer membranes	Hydrophobic
Stainless steel	Dispensing components	Stable wetting Hydrophobic Air bubble mitigation Low retention coatings Wear resistance

and are referred to as in a secondary plasma or remote plasma treatment. Part fixturing may be electrically isolated, electrically biased, or at the ground potential.

2. Process controller: The process controller is the module that monitors all the sensors and transducers, stores the program into nonvolatile memory, displays pertinent data, and assures program execution.
3. Plasma generator: A module whose function is to ionize a gas. This can be done through a high-frequency radio frequency (RF) power or a low-frequency electrical arc. The plasma generator provides energy required to ionize gas and maintain a plasma state.
4. Matching network: A module whose function is to assure efficient transfer of electrical energy from the plasma generator to the ionization chamber. Proper matching enables reproducible ionization of a gas ensuring long-term process reliability (*see* **Note 3**).
5. Electrodes: Located inside the ionization chamber and whose function is to deliver energy supplied from the generator.

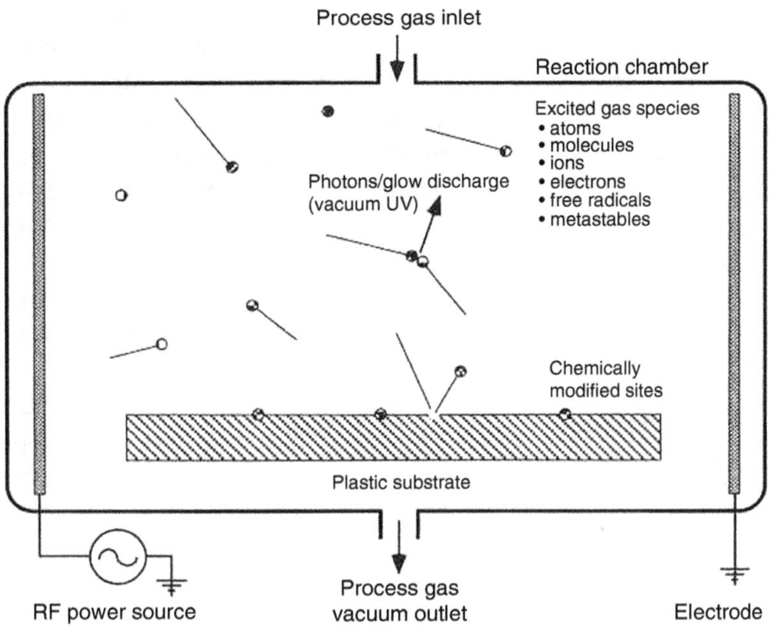

Fig. 3 Schematic of surface modification in a gas plasma reactor

Electrode configurations will vary depending on the type of plasma and the system throughput. A schematic example of a capacitively coupled vacuum chamber plasma is provided (*see* Fig. 3).

6. Gas/vapor flow controllers: Mass flow controller (MFC) or liquid flow controller (LFC) is a feedback control device designed to regulate gas or vapor flow rates to a desired set point. The MFC/LFC reproduces flow rate independent of temperature and independent of inlet or outlet pressure (*see* **Note 3**).
7. Pressure transducer: The transducer, also known as a capacitance manometer, is a sensor for measuring pressure in the ionization chamber. The output signal is read by the plasma process controller (*see* **Note 3**).
8. Vacuum pump: If the plasma is generated under a partial atmospheric pressure, then it will utilize a vacuum pump connected to the ionization chamber by a vacuum conduit. Plasma generated under atmospheric pressure does not require a vacuum pump. The function of a vacuum pump is to reduce the chamber to its operating pressure and expel used process gases. Pump selection is important for achieving adequate pumping capacity and gas dilutions. Polymer outgassing and materials with moisture uptake (where the moisture is not favorable for processing) require higher pumping capacity. Outgassed species are removed by the pump so that they do not become

undesirable products of ionization. Pump configurations vary and should be selected to ensure compatibility with the process chemistries (*see* **Notes 7** and **8**).

9. Vacuum valve: A pneumatically actuated solenoid valve is used to isolate the chamber from the vacuum pump. The valve is opened to pump down to base pressure. After purging, the vent valve is closed in order to return the ionization chamber back to atmospheric pressure.
10. Vacuum vent: After the process reaches completion, the ionization chamber must return to atmospheric pressure in order to unload and/or load parts. This process is referred to as venting and begins with the closing of the vacuum valve.
11. View port: The view port is a glass portal, sometimes with UV and RF filters, that enables viewing and monitoring of the plasma light emission inside the ionization chamber (*see* **Note 6**).
12. Base pressure: Base pressure refers to a set point whose function is to act as a pressure switch for a programmed process step. Equipment safety interlocks should be designed to ensure that the gases and the energy cannot be activated manually or automatically unless the system has reached the base pressure. A base pressure is selected that is lower than the processing pressure to ensure adequate evaluation of the chamber of outgassed species and moisture. Depending on the material, extended "pump downs" may be desirable to extend the outgassing step (*see* **Note 4**).
13. Step: A step is defined as a discreet duration of time when the plasma process gases flow with or without energy. A series of steps makes up the complete process. A process may be comprised of multiple steps, allowing an operator to execute a recipe with different steps in a single process. For example, a part may be cleaned in Step A with one chemistry and activated in Step B with another chemistry, without removing the samples or venting to atmosphere.

3 Methods

During a plasma process, energetic particles barrage a surface, causing chemical bonds to cleave and recombine. The dominant reaction with most polymer substrates occurs through hydrogen substitution along the polymer's aliphatic linkages. In the presence of oxygen, some hydrogen bonds are abstracted from the carbon backbone. The lost hydrogen may be substituted with oxygen to form polar compounds. The types of reactions and the concentrations of chemical compounds will depend on the plasma energy and the gas chemical composition. Gas or vapor combinations may be introduced into the plasma to influence the formation of desirable chemical compounds (i.e., hydroxy, carboxyl, amine). Producing a plasma with a lower gas pressure will have the effect of increasing a molecule's mean-free path. A plasma produced under partial pressure will have a longer mean-free path than a plasma produced under atmospheric pressure. A longer mean-free path allows an energetic particle to accelerate with greater ease under an electrical field. Collisions are less frequent, and the gas produces less heat due to the friction. Surface modifications under a partial vacuum pressure usually take place inside the region of primary plasma where there is greatest activity. Reactive species are more likely to access interstitial surfaces when placed within the primary plasma [2]. A plasma generated at higher pressure, such as under atmospheric conditions, produces more frequent collisions. The reaction time is much shorter as a result. However for most materials the processes are limited to a specific region within the secondary plasma zone where the plasma conditions are neither too weak nor too aggressive.

3.1 Plasma Process Design

A plasma process should be designed with consideration of the substrate material and the intended surface functionality. Not all materials will respond to a plasma treatment in the same way. What works on one material system will not always transfer to a different family or grade of materials. Therefore, the plasma process should be tailored and tested to a specific material. Even the fabrication of the material (e.g. machines or molded) can affect the outcome of the surface treatment. Whenever possible, a plasma treatment should be validated on the precise substrate material in its final form factor. The following section identifies a methodology for designing, testing, and increasing the throughput of a plasma process.

1. *Problem definition.* Identify the project goals with a detailed account of the desired surface characteristics. Surface chemistry alone may not address all the material challenges. Consider other factors that may influence your result: surface topography, environmental effects, storage conditions, confounding processing aids, packaging materials, and sterilization parameters. It may be possible to simultaneously address multiple surface property needs using a plasma process, e.g., roughen, functionalize, alter wettability (*see* **Notes 1** and **2**).
2. *Equipment selection.* There are various options and system configurations that should be reviewed at the start of the project development. For example, should the treatment be done in-line, selectively, or conformally? Note that the effects of a plasma process do not necessarily transfer from one equipment design to another. The results achieved using a small benchtop

chamber may not necessarily achieve an identical response in a reel-to-reel system, atmospheric plasma, or larger batch system. The electrode size, spacing, orientation, and power source can influence the concentrations of active species that may interact with the surface. Whenever possible, utilize an equipment configuration that most closely matches the final equipment format.

3. *Surface preparation.* Plasma interactions predominantly occur near the surface on the submicron or atomic length scale. Physical debris or contaminants confound the outcome of a surface modification. Some contaminants may originate as additives or pigments within the material bulk. Some undesirables on the surface may require additional bulk cleaning before the plasma processing. Only use solvents or detergents that are compatible with the material system. Additionally, the surface should be properly rinsed of residual cleaning agents and dehydrated (*see* **Note 1**).
4. *Plasma cleaning.* A short plasma cleaning step refers to a molecular etching intended to homogenize the surface before processing. This is generally the cleanest state that may be achieved on the surface. The plasma easily removes organic oligomers or low molecular weight molecules that were left over from the cleaning process. Cleaning is achieved through a deconstruction and removal of the carbon species via exposure to oxygen plasma. Gases, or a combination of gases, may be selected to create plasma species that are targeted toward a specific surface contaminant. Table 2 lists examples of gases and vapors that may be used within a partial atmosphere plasma reactor. A mixture of oxygen and tetrafluoromethane produces a plasma that is especially aggressive. Atmospheric plasma, on the other hand, primarily utilizes clean air or an oxygen-containing gas mixture. With judicious selection of process conditions, there should be no observable changes to the substrate following plasma cleaning. Overtreatment of polymeric materials could result in visible damage to the part. Some metals may tarnish if the treatment conditions are too aggressive. The effects of plasma cleaning may be reduced by lowering the plasma power or reducing the treatment time. Overtreatment may result in poor adhesions to the surface or poor treatment stability.
5. *Surface chemistry.* Plasma surface activation provides means of permanent covalent attachment of surface chemistry. Chemical species may be generated by flowing specific gases or vapors selections into the plasma (*see* Table 2). The correct combinations will result in repeatable and targeted functional species. An example of the types of species is shown in Fig. 4 and described in more detail below.

Table 2
Available gases and liquids for plasma chemistry applications

Gas	Liquids (create vapors)
Oxygen	Methanol
Argon	Water
Helium	Allylamine
Nitrogen	Ethylenediamine
Hydrogen	Acrylic acid
Nitrous oxide	Acetone
Carbon dioxide	Hydroxyethylmethacrylate
Air	Fluoroacrylates
Methane	Ethanol
Ethane	Toluene
Ethylene	Diaminopropane
Acetylene	Butylamine
Tetrafluoromethane	Gluteraldehyde
Hexafluoroethane	Hexamethyldisiloxane, octamethylcyclotetrasiloxane, tetramethylsilane
Hexafluoropropylene	Polyethylene glycols
	Diglymes Silanes (amino, carboxy, hydroxyl, mercapto, vinyl)

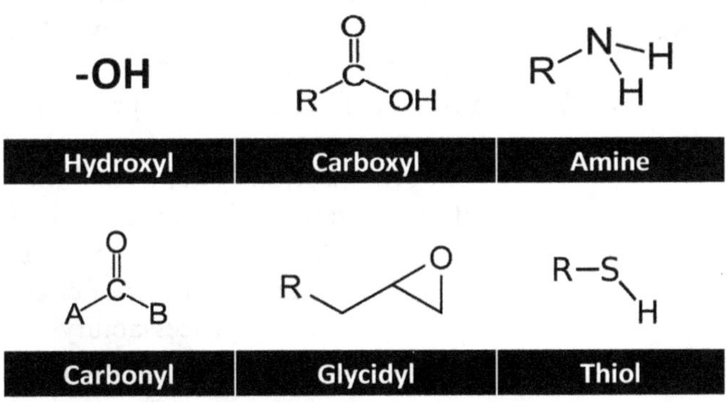

Fig. 4 Examples of plasma-created functional species

5.1 *Hydroxyl chemistry.* Gas plasma containing oxygen gas or alcohol vapor produces an adjustable concentration of surface hydroxyl groups. Hydroxyl groups link biological molecules through simple dehydration reactions. These also impart surface polarity, which may be utilized to enhance substrate wettability. The simplicity of generating hydroxyl groups makes it one of the most common plasma modifications.

5.2 *Carboxyl chemistry.* Gas plasma containing carbon dioxide gas, acetic acid vapor, or acrylic acid vapor is used to produce carboxyl groups. The carboxylic acid functionality is an integral part of amino acids and is especially useful in biomolecule capture. Under high pH, the carboxylic acid will disassociate to form a carboxylate ion. Carboxyl surfaces are hydrophilic, negatively charged, and capable of nonspecific binding.

5.3 *Amine chemistry.* Gas plasma containing ammonia gas, diamine vapor, or allylamine vapor produces primary, secondary, and tertiary amine. Primary amines donate hydrogen atoms to form hydrogen bonds. Amines are commonly employed in the capture, immobilization, and retention of biological molecules such as proteins, lipids, and antibodies.

5.4 *Custom chemistry.* Plasma processing enables many other possibilities in surface chemistry. Judicious selection of the gas and vapor species enables an operator to design and customize their surface chemistry to include deposition of thin film coatings by the process known as plasma-enhanced chemical vapor deposition (PECVD). Example PECVD surfaces include hydrophobic and low-retention coatings that may be used for dispensing components (*see* Table 1).

3.2 Process Development

The primary variables to study in a plasma process development are gas and/or liquid vapor flow, plasma power, and exposure time. Keep in mind that polymers are complex systems and may express different affinities or susceptibilities to specific plasma gas compositions or exposure times (*see* **Note 1**). A plasma recipe that works with one material system may not necessarily work the same on a different material system. A basic methodology for determining a successful plasma condition is to start with a screening of plasma chemistry. Then, the result of chemistry may be further optimized for uniformity or processability. The following are some steps to consider in designing a plasma process:

1. *Chemistry screening.* Process development begins with a screening of different plasma gas and vapor chemistries (*see* Table 2). The objective would be to determine which plasma chemistries provide the most notable improvement in the surface properties. A good plasma candidate should produce a substantial positive signal. The best process candidates may be studied for further development and optimization. For example, a

development for achieving surface amine may start with a comparison of plasma using ammonia, allylamine, or aminopropyltrimethoxysilane. Each of these processes is expected to deliver amine to the surface; however, the chemistry screening will determine which plasma chemistries have the best response for the application. The ammonia process might yield a sufficient density of amine but that amine might not be accessible because the amine forms as a pendant side group too close to the polymer backbone. The allylamine, on the other hand, may locate on either a short chain or a branched structure that makes it more accessible at the surface (*see* Fig. 4). The plasma chemistry screening tests how susceptible the material is to the plasma treatment.

2. *Process optimization.* Other properties of the surface may be optimized once a good plasma chemistry benchmark is established. The film thickness may be modified or the degree of fluid wettability. Process improvements are made by investigating the primary and secondary process parameters. The primary variables are typically power and plasma exposure time. The plasma power and process time will usually have a large impact on the density of surface functional groups as well as on the stability of the treatment. If the plasma power becomes too strong, then the gas species fragment leads to poor chemistry selectivity and ultimately lower performance. At low power, the concentration of surface functional groups may be too low. If the treatment time is too short, then the concentration of functional groups may be too low. The secondary plasma power process variables include pulse frequency and duty cycle. These variables may be used to influence branching, fragmentation, surface roughness, and surface temperature.
3. *Gas or vapor flow.* The concentration and composition of the plasma chemistry affect the generation of surface chemistry. Flow management systems (such as MFCs and LFCs described in Subheading 2) will regulate and ideally monitor flow into the plasma chamber. Vapor chemistries are typically accompanied by a co-process gas. They will assist in sustaining a uniform glow. Inert gases such as argon and helium make good candidates for co-process gas because they do not form chemical bonds with the primary gas chemistry. Oxygen-containing compounds will have a tendency to oxidize other species within the chamber. This may be undesirable in an example where the projected amine would be converted to an undesired nitrogen oxide.
4. *Chamber pressure.* The process pressure affects the process temperature and the plasma density. The process pressure usually results from the flow of species into the reactor chamber. Regulating pressure independent of flow is not necessary;

however, some process chemists favor controlling the chamber pressure by adjusting pump impedance at the vacuum port using a throttle valve (*see* **Note 5**). Higher pressure plasma processes generally contain more thermal energy, whereas lower pressure plasma contains more energetic species.

5. *Continuous* vs. *pulsed power.* Plasma power is one of the notable influences on the surface modification or polymerizations. Some plasma generators come with the ability to pulse the power and employ a duty cycle. Pulse and duty cycle enable an operator to reduce plasma temperature. This is especially critical in the modification of thermally sensitive materials. Plasma temperatures may also introduce surface stress, which could negatively impact treatment stability in some material systems. Pulse and duty cycle are valuable controls for plasma film formation. It helps by regulating kinetic events within the polymerization reaction. The power duty cycles have the ability to influence coating thickness, coating chemistry, and coating density.
6. *Interstitial modification.* Partial vacuum plasma processing is advantageous for modifying the interstices of porous media, in addition to the visible outer surfaces. This technique is ideal for changing the wicking and binding properties of sintered, woven, nonwoven, and membrane materials that are commonly used for filtration (*see* Table 1). As described herein, the equipment used creates primary plasma. In primary plasma, there is a greater mean-free path of the particles before a collision. This sustained energy is ideal for modifying the interstices of porous media or for use inside complex nanoscale vials or channels. With atmospheric processes, on the other hand, the mean-free path is short when the species are out of the direct energy field, thus the treatment zone is limited. The first step of the process, where modification in either the inner dimensions or interstices is required, requires conditions to ensure that the polymer is outgassed. In all plasma processes, the reaction occurs between the plasma and surface. Removing barriers to the surface is critical for a meaningful modification. In a vacuum process, the first step is the pump-down phase, which is further described in Subheading 3.3. During an outgassing step, extended time for the pump down as well as a lower base pressure is important (*see* **Note 4**). Following the pump down, the plasma may be introduced continuously or via a multiple-step process containing numerous pump downs. An example of the wicking and wetting effect of a high-density polyethylene, following an oxygen plasma process, is shown in Fig. 5. With judicious selection of plasma processing parameters, surface stability may persist for many years of storage prior to the final end use.

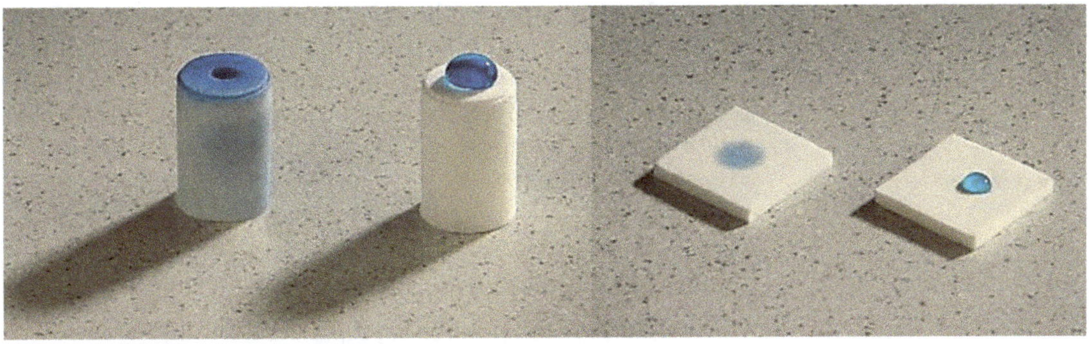

Fig. 5 Interstitial modification of porous media for enhanced wicking

3.3 Plasma Treatment Under Partial Pressure

Once a study has been outlined, the steps to partial vacuum processing are listed in simple terms as follows:

1. The components, to be treated, are placed in the reactor chamber, which is vacated via a pumping system, to a base pressure that may range from 25 to 500 mTorr (*see* **Note 4**). During this initial processing step, the vacuum pump isolation valve is in the open position.
2. Process gas(es) and/or liquid vapors are introduced into the chamber at a desired flow rate to attain a specific pressure (*see* **Note 3**).
3. RF energy (as used in the Plasma Science, PTS, and Plasmatreat partial vacuum systems) is supplied to electrodes within the chamber and excites the vapors/gas(es) into a plasma (*see* **Note 3**).
4. During processing, pressure is continuously monitored as an indicator of chemistry uniformity (*see* **Notes 6–8**).
5. While under vacuum, a sequence of multiple steps may be conducted as part of a single recipe. Between each step and after the last step, the chamber is pumped down to the base pressure to evacuate gases from the previous process step (*see* **Notes 4** and **8**).
6. Following completion of the last step, there is a venting cycle in either room air or a neutral gas such as nitrogen to bring the chamber to atmosphere. The venting speed is controlled to minimize disruption to fragile components if required.
7. The parts are removed and packaged or transferred to a subsequent process (*see* **Note 2**).

3.4 Plasma Treatment Using Atmospheric Plasma System

1. The components to be treated are placed securely on a stage. Either the stage will provide motion to the part or the part may remain fixed while the jet head is moved. Movement speeds may vary from 10 to 1000 mm/s.

2. Process gas(es) and/or liquid vapors are delivered to the plasma nozzle. Time is allowed for the flow to equilibrate. Properly engineered exhausts must be implemented to ensure proper management of the unused gas effluents, which may require particle filtration and an external exhaust port to the an exterior environment.
3. Plasma discharge is initiated, and the plume is scanned across the sample area with precise control over the speed and the gap between sample and plasma nozzle.
4. During the process, gas flows, jet pressures, and plasma power are precisely monitored.
5. Generally, the plasma treatment may be applied in one pass; however, large surface areas or thicker nanocoatings may be achieved by performing multiple passes over the sample area. Higher speeds should be maintained for thermally sensitive materials due to the buildup of temperature while the plasma is being generated.
6. The plasma outlet should always remain unobstructed. After the plasma processing, a preventative measure is made to clean the nozzle tip using a brush to avoid building up of polymerizable species. If there is no polymerization, then the preventative cleaning step is not required.
7. The plasma-treated parts are removed and then packaged or transferred to a subsequent process (*see* **Note 2**).

3.5 Evaluation of Plasma-Treated Surface

Determining a validation method is equality important in designing the plasma process. It is critical to validate the surface with a test that is representative or that closely correlates with the end application. A few popular techniques for quantifying and qualifying the effects of a plasma surface modification are described herein.

1. *Surface energy.* An alteration of the surface chemistry will usually change the surface tension. Contact angle and dyne test inks provide a means for rapid examination of surface tension. Water will generally form lower contact angles, or spread out, on polar surfaces. The contact angles are measured between the substrate and the tangent where the droplet meets the surface. Only use distilled water when making a water contact angle measurement. Low-surface wetting angles usually correlate with an abundance of oxygen functionality. Another way to measure surface tension is with a dyne test ink. These fluids usually come in kits containing different surface tension. A dyne test fluid must have a lower surface energy relative to the surface in order to wet out. If the fluid has a higher surface energy, then it will bead up or retract from the surface. By observing the wetting behavior of various fluids, one may ascertain the surface tension. The surface dyne method and the water contact angle method provide a facile methodology for probing the plasma-treated surface. Note that

these methods do not necessarily discriminate chemical functionality. The author recommends testing functionality or using analytical methods for characterizing surface chemistry before establishing process quality controls using test fluids such as water or dyne test ink.

2. *Chemical analysis*. There are numerous analytical tools that are able to determine a surface's chemical state or composition. Some examples include X-ray photoelectron spectroscopy (XPS), nuclear magnetic resonance (NMR), time-of-flight secondary ion mass spectrometry (TOF SIMS), and energy-dispersive spectroscopy (EDS). These and other similar methods are useful in attributing surface effects to the underlying surface chemistry. Unfortunately, many of these methods are not readily available due to cost or accessibility to trained specialists.
3. *Markers/stains*. Staining and fluorescent probe methods are especially useful in evaluating the surface. Testing is highly functional and typically relevant to the application. A successful conjugation will continue to show good retention of surface chemistry even after washing. Evaluate the surface for good selectivity and a low coefficient of variation. Low-chemistry retention could be an indication of a poor surface treatment or a problem with surface contamination. A high coefficient of variation could indicate poor chamber disposition or a confounding surface contaminant.
4. *Stability and accelerated aging*. Treatment stability should be assessed for every application. Many of the reactive surface chemistries are sensitive to environmental effects. These effects might include moisture, temperature, light, exposure to solvents and oxidation. If the parts need to be stored, use a clean packaging material. Many plastic bags contain plasticizer or antiblocking agents that will confound a plasma surface treatment (*see* **Notes 1** and **2**). The author uses heat-sealable forensics bags for short-term storage. Some polymer substrates are especially prone to aging. In general, the surface treatment will diminish quickly on low durometer elastomers and on materials with low glass transition temperatures where the polymer chains are mobile near ambient conditions. Mobile chains lose surface treatment due to reorientation and other kinetics. Accelerated aging may be useful in simulating the long-term stability.

4 Notes

When designing a plasma process, take a holistic consideration of the equipment design, product material, process recipe and subsequent handling environment. Once a plasma chemistry has been achieved, if you observe irregular results (uneven treatment), consider further examination of the surface cleaning or other preparation step. You may also want to ensure repeatable substrate placement.

1. Material variability may impact the performance of plasma-treated surfaces. Therefore, it is pertinent to define your material grade and material manufacturer before starting plasma surface treatment. A change in polymers and/or fabrication method may impact the plasma process. Some material variables include.
   (a) Additives (stabilizers, pigments, nucleating agents, plasticizers) and the propensity for migration of additives.
   (b) Stresses or propensity for molecular rotation.
   (c) Finishes/roughness.
   (d) Mold release materials.
   (e) Machining debris.
   (f) Moisture uptake.
   (g) Cleanliness (and how is the substrate cleaned prior to plasma).
   (h) Susceptibility to oxidation.
2. Subsequent steps post process can impact the performance of the plasma-treated surface. These may include.
   (a) Adhesive techniques (bonding, heat sealing, ultrasonic welding).
   (b) Chemical/solvent exposure.
   (c) Cleaning and/or sterilization processes.
   (d) Secondary contamination from storage or handling post-plasma.
   (e) Monitoring selection of glove type and compatibility of the glove materials with chemicals.
   (f) Storage selection and proper packaging material are important to limit plasma-treated surface contamination.
   (g) Temperature, light, or environmental exposure.
3. Power and gas flow deficiencies should be monitored. If the equipment used is manually controlled or lacks capabilities to define set points for a given variable, carefully monitor the output device. Monitoring with a secondary calibrated tool may be required. Be cognizant of reflected or lost power where the set point energy is not actually delivered.
4. When designing a process, choosing the base pressure is important. Too high of a base pressure may impact the outgassing process. Base pressure should always be below your operating

pressure to ensure outgassing of the part and chamber. The chamber may also absorb moisture during inoperation. Carefully monitor the time to reach base pressure in an empty chamber and compare this time to when a chamber has parts in it or after long periods of unuse.

5. Once the process has been developed and fixturing optimized, take great care to observe and to monitor the pressure. A change in the pressure is a signal that a component may not be working per an intended specification and/or there is chamber or part contamination. Note that some plasma equipment includes a throttle valve. A throttle valve "throttles" the pumping capacity of the chamber, which modifies the pressure of the species inside the chamber. Equipment supplied with automated throttle valves work by changing this throttle (thus vacuum conductance) to meet a pressure set point. This maintains a constant pressure in the chamber by adjusting the vacuum conductance. When using a throttle valve, be sure to check the health of your pumping system and flow controllers as the condition may be masked due to the automatic adjustment.
6. One of the simplest ways to monitor the plasma process is by observing the plasma color. A change in color may be indicative of gas flow issues, air leaks, excessive outgassing, or an unclean chamber.
7. The nature of the vacuum plasma process is to introduce new chemical species to surfaces. These species may deposit on the inner surfaces of the chamber, thus design of a chamber cleaning protocol helps to ensure reproducible results. This includes manually wiping the chamber of dust and cleaning the exposed surface between the chamber and isolation valve, and a periodic plasma cleaning process to remove organic build up.
8. Regularly perform maintenance checks to ensure that there are no leaks due to loose fittings or worn sealing surfaces. It is advised to maintain a log of daily (or weekly) checks to monitor both the pump-down speed and leak-up rate of the chamber with and without parts. A drift in either of these conditions is an immediate warning of either a system issue or materials issue.

References

1. Plasma Science and Plasma Technology Systems Equipment Operations Manual Glossary Section, 1997 through 2015
2. Plasma Science, Technical Note 7, October 1989

Chapter 7

Interference in ELISA

Robert S. Matson

Abstract

ELISA is a well-established technique used worldwide to quantify analytes present in a diverse milieu of biological samplings. It is especially important to clinicians who rely on the accuracy and precision of the test to administer patient care. Those results are to be held with great scrutiny since the assay is subject to error caused by interfering substances found in the sample matrix. In this chapter, we examine the nature of such interferences and discuss approaches to identify and offer remedies to remove the interference and validate the assay.

Key words Interference, HAAA, HAMA, Autoantibody, Heterophile, Biotin, HRP, ALP, Substrate inhibition, Hook effect, False positive, False negative, Endogenous, Exogenous, Analytical error, Spike, Recovery, Dilution linearity, Matrix effects

1 Introduction

After carefully designing ELISA, establishing the conditions for antibody coating and blocking, and running standards in a defined buffer, the real work begins! You spike your calibrated standards into serum and remeasure. Alas, in our case procalcitonin (PCT)-recovered values were found to be much lower than expected (Table 1). Something in the serum was adversely affecting the accuracy of the test. That is interference!

Both false-negative and false-positive indications can occur depending upon the sample matrix. For example, in an experiment where we spiked human chorionic gonadotropin (hCG) into urine, a negative interference occurred while spikes into plasma resulted in a false positive (Fig. 1). The difficulty with interference is recognizing that it is occurring. One must take measures to ascertain whether or (hopefully) not there is a problem. In this chapter, we discuss approaches to determining the nature of the interference and then possible remedies.

Robert S. Matson (ed.), *ELISA: Methods and Protocols*, Methods in Molecular Biology, vol. 2612,
https://doi.org/10.1007/978-1-0716-2903-1_7,

Table 1
Human procalcitonin assay interference from spike samples

Sample	Mean intensity	Observed concentration (pg/mL)	Expected concentration (pg/mL)	Recovery (%)	Dilution
	2121	17,862.60	20,000.00	**89**	Undiluted
	1863	7461.90	6666.70	**112**	1/3
	1165	2D76.7	2222.20	**93**	1/9
	596	766.7	740.7	**104**	1/27
	242	269.8	246.9	**109**	1/81
	72	64.5	82.3	**78**	1/243
	36	18.8	27.4	**68**	1/729
	18	OR-low	0		Blank
Serum spike	522	651.5	6666.70	**10**	1/3
Serum spike	232	257.5	2222.20	**12**	1/9

Obs observed concentration of analyte from standard curve fit, *Exp* expected or experimental input concentration. Recovery (%) = Obs − Exp/Exp × 100

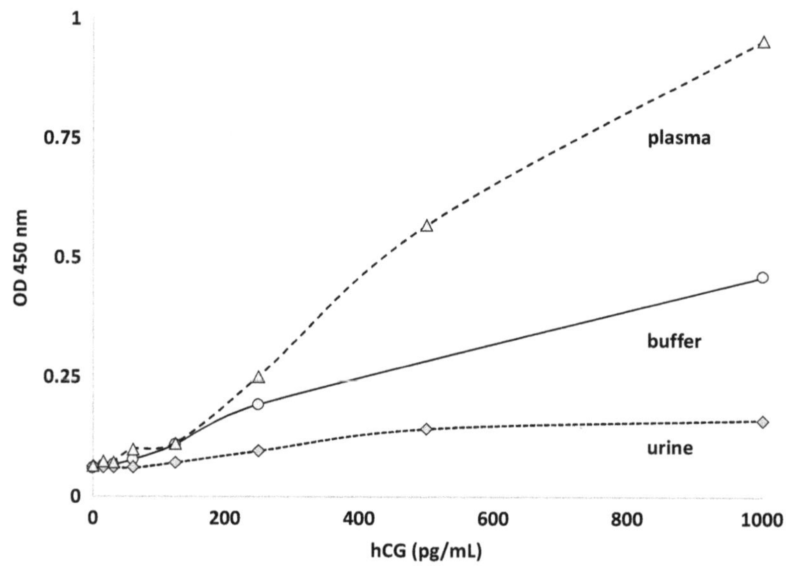

Fig. 1 Positive and negative interferences during hCG ELISA. Human chorionic gonadotropin (hCG) spiked into urine diluent (1:1) obtained from an adult female exhibited a false-negative interference relative to the analytical standard curve where hCG was serially diluted into 1% BSA–PBS, pH 7.4 diluent only. The hCG spiked into plasma diluent (1:1) resulted in a high false positive

2 Interference or Not?

Interference is the result of a substance(s) when present in a sample that leads to an inaccurate measure of the true value of an analyte [1]. What kind of interferences are associated with immunoassays and in particular ELISA? There are multiple scenarios to consider. Ward et al. [2] describe with several good clinical examples the kinds of interferences that are usually associated with the immunoassay. These may be generally characterized as either "analyte-independent" or "analyte-dependent." Hemolysis and lipemia; high doses of ingested biotin [3] that saturate streptavidin signal reporter conjugate or the presence of high levels of hCG that give a false negative via the hook effect [4] would be examples of analyte-independent interferences. That is, they do not directly involve a perturbation or disturbance in the measurement of analyte.

Here, we examine those analyte-dependent interferences that are primarily cross-reactive and competitive with the components of ELISA. At the antibody level (*see* Fig. 2), this may involve the masking of antibody binding, sequestering of available antigen, or competition with exogenous or endogenous antibody [5]. The presence of heterophile antibodies and autoantibodies (e.g., rheumatoid factor) in patient sera are well-known sources of interference for ELISA [6]. Endogenous levels of human antianimal

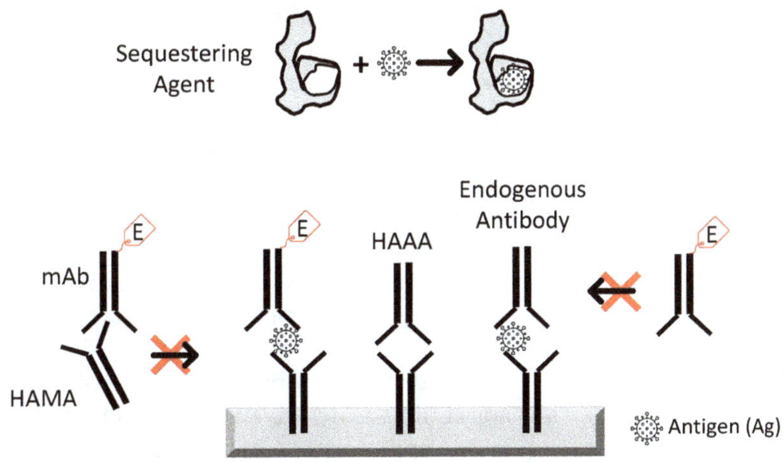

Fig. 2 Interference at the antibody level. The sandwich ELISA is prone to interference from different binding agents found in biological matrices such as plasma, serum, or urine. Sequestering agents such as human serum albumin or other endogenous binding proteins could remove antigen (e.g., therapeutic drug, hormones), thereby preventing binding to either capture or reporter antibody. Complement fixation of the Fc region of antibody may sterically hinder antigen binding. Heterophilic antibodies and human antianimal antibodies (HAAA) such as mouse monoclonal antibody (HAMA) are well-known interferences

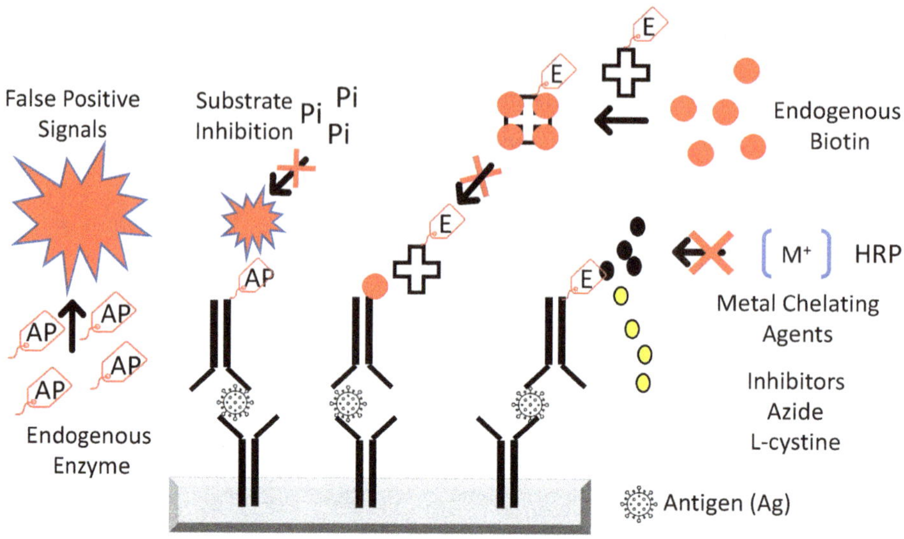

Fig. 3 Interference at the enzyme level. The presence of metal chelating agents found in reagents or the sample matrix can inhibit enzymatic activity by removal of metal ion cofactors from an enzyme. Horseradish peroxidase requires Fe-heme and Ca^{2+}, while alkaline phosphatase utilizes Zn^{2+} and Mg^{2+}. Ascorbic acid, azide, and thiols such as L-cystine can be potent inhibitors of HRP activity. Endogenous biotin overload from vitamin supplements challenges the utility of streptavidin–biotin reporter-based ELISAs. The presence of endogenous alkaline phosphatase in serum can result in false positives, while excess inorganic phosphates can cause a negative interference with the enzyme

antibodies (HAAAs), especially human antimurine antibody (HAMA), are also prevalent and perhaps more problematic [7]. The sandwich immunoassay can be significantly affected by bridging heterophile or HAAA with the secondary and/or capture antibody. At the enzyme reporter level (*see* Fig. 3), there may be endogenous enzyme (e.g., serum alkaline phosphatase) or various enzyme inhibitors such as metal chelating agents, and redox chemical species such as peroxide or azide, which will reduce the activity of horseradish peroxidase. The presence of inorganic phosphates in serum is a source of substrate inhibition of alkaline phosphatase [8], while substances found in urine are known to suppress horseradish peroxidase activity [9]. As discussed earlier, an important consideration currently looming is the use of the streptavidin–biotin system for clinical assays given that high levels of endogenous biotin from food supplements are now being seen.

3 Fixing the Problem

It is prudent to first examine the assay for possible interference. The simplest step is to serially dilute the sample and rerun ELISA. If the activity rises upon dilution, it is likely that either there is an interfering substance in the sample (antibody level) or a hook effect

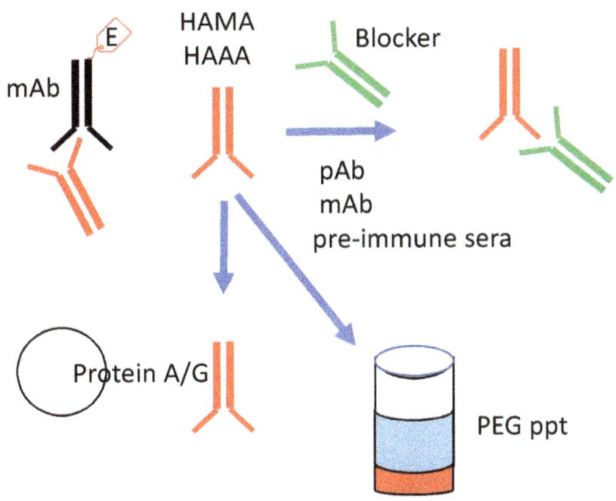

Fig. 4 Reduction in human antianimal antibody interference. The presence of HAMA/HAAA in samples can lead to severe loss in assay sensitivity and an underestimate (low recovery) of the analyte concentration. There are three main approaches to reduce the interference. The simplest means is to add a blocking agent that will bind to the heterophile antibody such as an antispecies antibody or an excess of nonspecific mouse monoclonal to compete out the HAMA or use preimmune sera. Other approaches involve removal of the offending antibody, for example, capture to a Protein A/G solid phase or precipitation using polyethylene glycols (PEGs)

(enzyme level) is present. There is no single approach that will guarantee complete success in resolving the interference, but there are means to mitigate most issues [10]. Assuming that interference is indicated, the following approaches may offer remedy: sample dilution, addition of a heterophile blocker, addition of pre-immune sera or nonspecific mouse monoclonal antibody, or lastly removal of the interfering antibody by physical means such as adsorption to an affinity capture solid phase (e.g., Protein A/G gel) or a precipitation method (*see* Fig. 4). Moving to a different ELISA format may also be warranted. First, however, test the sample using another vendor's ELISA or a different diluent just in case the interference is associated with the analytical reagents. Otherwise, if HAMA cannot adequately be reduced using blocking agents try using a different species other than mouse monoclonal such as rat monoclonal, chicken polyclonal, an antibody fragment (Fab), nanobody (nb), or aptamer.

The impact of interference in ELISA is most critical in the clinical laboratory where false-positive and false-negative determinations are not to be tolerated. Wauthier et al. [11] offer an algorithm to systematically sort out sources of interference and methodology to apply in remedy. Essentially, the approach begins with a starting point "suspicion of an interference" from the laboratory that is based upon previous experience or clinical outcomes (I.e., this case is unusual or inconsistent with previous tests). The next step is to decide whether there were any "pre-analytical errors"

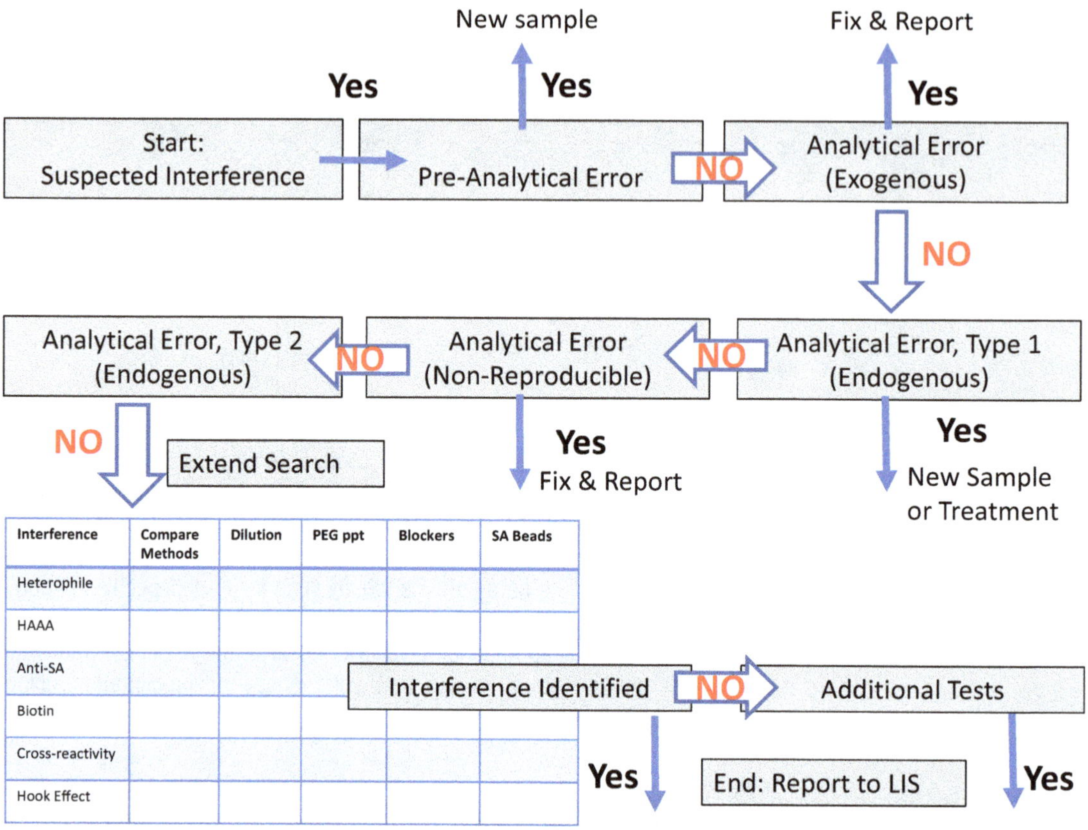

Fig. 5 Algorithm workflow for identification and resolution of interference. The diagram provides a systematic approach to resolve interference in an immunoassay in the clinical laboratory setting. (Redrawn from Wauthier et al. [11] with modification)

such as with incorrect sample collection or preparation (*see* Fig. 5). If yes, prepare a new sample. If no, look for analytical errors due to, for example, instrument calibration ("exogenous analytical error"); hemolysis ("endogenous analytical error, type 1") or carryover, pipetting error, etc. ("nonreproducible") or presence of heterophilic antibody, autoantibodies, biotin, etc. ("endogenous, type 2") according to the algorithm. From there, the laboratory can choose to perform various procedures to identify and remove the interference as discussed.

Kragstrup et al. [12] have provided a series of "essential" steps to be taken to validate a sandwich ELISA free of interferences:

1. False positives: Coat wells with capture antibody (antigen-specific binding) and an isotype control antibody (no specific antigen binding). Prepare a biological sample matrix (e.g., plasma) for testing with and without a blocking agent (e.g., heterophile blocking reagent) and compare binding signals. True-positive signal should be present with the wells containing capture

antibody but not the isotype antibody wells. If signal is present on isotype antibody from unblocked sample, a false positive is indicated. If the signal is greatly reduced or missing on the blocked sample, then the interference is resolved.

2. False negatives: Conduct a spike-recovery test using a known concentration of antigen doped into the sample matrix. A low recovery of the antigen indicates interference, leading to a false-negative result. Add a blocking agent to the doped sample and determine recovery. If the antigen level is now accurately predicted within 80–120% of the value from the standard curve, then the interference is resolved by addition of the blocking agent.
3. Matrix effects: Serial dilute the sample matrix and perform ELISA with and without blocking agents in the matrix as before. A sample free of interference should show linearity with dilution; otherwise, an interference is suspected. If linearity is recovered by the addition of a blocker, then the interference is resolved.

Spike-Recovery Method

Andreasson et al. [13] have provided a definitive guide to immunoassay methods validation that includes succinct explanations of validation parameters and standard operating procedures (SOPs) for their determination. As noted, interference can be evaluated using the spike-recovery method. The purpose of the test is to determine whether the value obtained for analyte (antigen) in the sample matrix is in good agreement (recovered) with that reported for analytical standards obtained using the calibrator dose–response curve. Below is a slightly modified version of their procedure for determining recovery:

1. Using analytical standards (calibrators), prepare a dose–response curve in diluent. Determine the linear dynamic range and then subdivide the curve into low, medium, and high regions (Fig. 6). For example, low at 25%, medium at 50%, and high at 75% of the upper limit of quantitation (ULOQ). The ULOQ represents the highest mean dose–response point at ≤20% CV within the linear dynamic range (i.e., prior to signal saturation).
2. Obtain a sample for study (e.g., test serum) and aliquot into four equal volumes. Spike three of the four aliquots with low, medium, and high standards, respectively. Spike the fourth aliquot with a like volume of diluent. This serves to correct for any dilution effect.
3. Prepare a standard curve as in **step 1**. Include the spiked samples from **step 2** in the ELISA analysis with the standards.

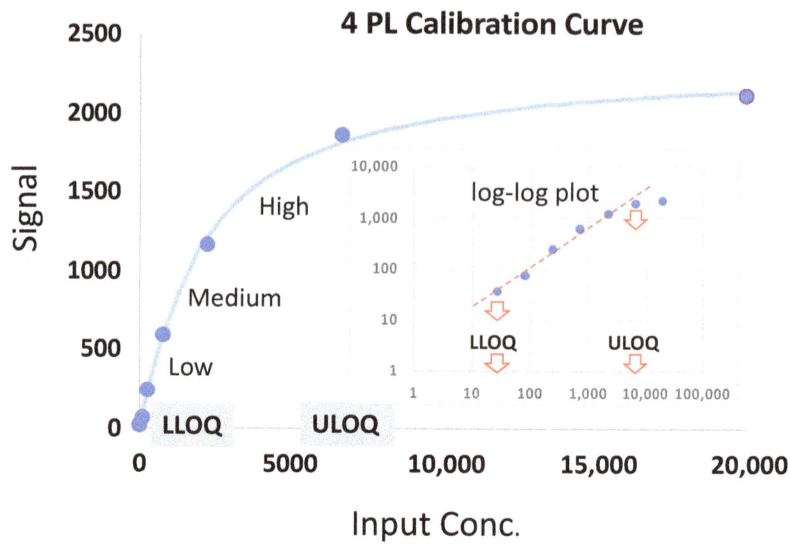

Fig. 6 Spike-recovery method from the analytical calibration curve. Standards (calibrators) are first evaluated for dose–response in diluent. The upper limit of quantification (ULOQ) and lower limit of quantification (LLOQ) are established from replicates with a cutoff at <20% CV within the linear range of the assay. A log (mean signal) versus log (input concentration) can be used to determine the linear range. Next, calibrators are selected within the ULOQ-LLOQ range at high, medium, and low concentration and spiked into the sample matrix. ELISA is performed with the standards and the spiked calibrators; and the recovery determined for each spiked sample. Once established, the process can be repeated with blocking agents to reduce interference

4. Calculate the recovery for each sample (*see* Table 1):
   (a) % Recovery = $[\text{Spike}]_{\text{Obs}} - [\text{Standard}]_{\text{Obs}} / [\text{Spike}]_{\text{Exp}} \times 100$

   where Obs is the observed (calculated) concentration [x] from the standard curve fit, while Exp is the expected or experimental (input) concentration.

Dilution Linearity Method

It is important that samples are analyzed within the working range of the assay's calibration curve. Otherwise, the sample cannot be accurately quantified. This often requires dilution of the sample. Provided the absence of interference, the serial dilution of the sample should yield a linear dose–response once dilution below the ULOQ is reached. As discussed earlier, this is recommended as the first approach to be taken in validating an ELISA especially to determine whether there is a hook effect [13].

1. Spike samples with calibrators at concentrations greater than the ULOQ.

2. Serially dilute the spiked samples to less than the LLOQ.
3. Perform the ELISA with replicates of the diluted spiked samples.
4. Calculate the mean dose and recovery for those diluted samples that fall within the linear dynamic range. Select dilutions that show recovery within 80–120% of the input concentration.
5. Plot signal versus dilution factor. Observe whether a hook effect occurs at high-analyte concentration. Perform sample dilution as needed to work outside that range.

References

1. Tate J, Ward G (2004) Interferences in immunoassay. Clin Biochem Rev 25:105–120
2. Ward G, Simpson A, Boscato L, Hickman PE et al (2017) The investigation of interferences in immunoassay. Clin Biochem 50:1306–1311
3. Giovanella L (2019) The impact of biotin interference on laboratory testing and patient diagnosis in clinical practice. Int J Pharm 4(1): 1–4
4. Gronowski AM, Cervinski M, Stenman U-H, Woodworth A et al (2009) False-negative results in point-of-care qualitative human chorionic gonadotropin (hCG) devices due to excess hCG β Core fragment. Clin Chem 55(78):1389–1394
5. Dodig S (2009) Interferences in quantitative immunochemical methods. Biochem Med 19(1):50–62
6. Ward G, McKinnon L, Badrick T et al (1997) Heterophilic antibodies remain a problem for the immunoassay laboratory. Am J Clin Pathol 108:417–421
7. Gace M, Prskalo ZS, Dobrijevic S, Mayer L (2015) Most common interferences in immunoassays. Libri Oncol 43(1–3):23–27
8. Coburn SP, Mahuren JD, Jain M et al (1998) Alkaline phosphatase (EC 3.1.3.1) in serum is inhibited by physiological concentrations of inorganic phosphate. J Clin Endocrinol Metab 83(11):3951–3957
9. Campbell KL, Lopresti M, Lukas W (2014) Inhibition of horseradish peroxidase (HRP) by a nonhydrophobic component of urine: a caution for immunoassays. Open Clin Chem J 7:1–7
10. Schiettecatte J, Anckaert E, Smitz J (2012) Interferences in immunoassays. In: Chiu NHL (ed) Advances in immunoassay technology. InTech. isbn: 978-953-51-0440-7. http://www.intechopen.com/books/advances-in-immunoassay-technology/interference-in-immunoassays
11. Wauthier L, Plebani M, Favresse J (2020) Interferences in immunoassays: review and practical algorithm. Clin Chem Lab Med 60(6):808–820
12. Kragstrup TW, Vorup-Jensen T, Deleuran B, Hvid M (2013) A simple set of validation steps identifies and removes false results in a sandwich enzyme- linked immunosorbent assay caused by anti- animal IgG antibodies in plasma from arthritis patients. Springerplus 2(263):1–10. http://www.springerplus.com/content/2/1/263
13. Andreasson U, Perret-Liaudet A, van Waalwijk van Doorn LJC et al (2015) A practical guide to immunoassay method validation. Front Neurol 6(179):1–8

Chapter 8

Determination of Proinflammatory and Antiinflammatory Cytokines by ELISA Technique

Osmany Blanco Muñoz

Abstract

The main objective of this chapter is to determine inflammatory and anti-inflammatory cytokines in patients with or without preeclampsia by means of the enzyme-linked immunosorbent assay (ELISA) technique. In this chapter, 16 cell cultures were obtained from different patients admitted to the hospital with term vaginal delivery or cesarean section. Here, we describe the ability to quantify the presence of cytokines in cell culture supernatants. The supernatants of the cell cultures were collected and concentrated. The concentration of IL-6 and VEGF-R1 was measured by ELISA to determine the prevalence of alterations in the samples studied. We observed that the sensitivity of the kit allowed the detection of several cytokines in a range between 2 and 200 pg/mL. The test was performed using the ELISpot method (5×), which allowed a higher level of precision.

Key words Direct ELISA, Indirect ELISA, Sandwich ELISA, Blocking ELISA, Antibodies, Spectrophotometers, ELISpot

1 Introduction

The enzyme-linked immunosorbent assay (ELISA) technique is one of the best validated and most widely used assays for biomarker detection and diagnosis of diseases affecting the world's population [1]. Thus, it is used to determine the antigen or antibody concentration of a small molecule in complex sample mixtures on a routine basis in research.

This technique was first described by Yalow and Berson (1959) [2]. In the need to detect molecules in complex substances, these researchers used antibodies bound to radioactive compounds [2]; but due to the risks of using radioactive substances, the method was modified [3].

Consequently, they looked for other possibilities and determined laboratory tests where they showed that some enzyme–substrate combinations could change the color of the reaction

Robert S. Matson (ed.), *ELISA: Methods and Protocols*, Methods in Molecular Biology, vol. 2612,
https://doi.org/10.1007/978-1-0716-2903-1_8,

and of the medium, and that this change was quantifiable. This led to the search for enzyme–substrate combinations that could detect the molecules of interest.

Years later, two research groups in Europe were responsible for modifying and establishing the ELISA method we know today [4]. The experiment was carried out on rabbit serum where the amount of IgG was detected.

Currently, this method has had greater relevance in the coronavirus pandemic, where molecular tests are considered the reference standard for the diagnosis of coronavirus disease (COVID-19). However, serological and immunological tests may be useful in specific settings [5]. ELISA has been one of the most popular tests among laboratorians as it has higher sensitivities for detecting the presence of antibodies to SARS-CoV-2 [6].

Types of ELISAs

ELISA allows qualitative and quantitative detection of a substance, which is usually an antigen, immobilized on a microplate or directly bound to a capture antibody.

There are four main types of ELISAs: direct ELISA, indirect ELISA, sandwich ELISA, and blocking ELISA.

Direct ELISA consists of coating the plate with the specific antibody against the antigen to be determined. Then, the sample with the antigen of interest is added and the secondary antibody labeled with the enzyme is added. This complex will give a colored solution when a substrate is added that will react with the enzyme, which will be quantified in a spectrophotometer [7].

Indirect ELISA is similar to direct ELISA in that it consists of the detection of two antibodies that will be found together and are called primary and secondary. The latter will be conjugated with an enzyme that binds to the unlabeled primary antibody, which in turn is attached to the antigen of interest. Subsequently, a substrate is added that, upon reacting with the enzyme containing the secondary antibody, will produce a color change or a light signal that correlates with the amount of antigen present in the original sample [8, 9]. In this way, the molecule of interest is determined and measured in a spectrophotometer (Fig. 1).

In sandwich ELISA, a solid support plate is added to an antigen that is incubated for a specific time to bind strongly, then we call this sensitization to the plate, then on the top is added the sample that has the molecule (antibody) to be determined. If there are specific antibodies in the serum, the formation of an antigen–antibody complex will occur, which will later be detected by the addition of a second antibody directed against the complex formed (Fig. 1).

The other method is the blocking or competitive ELISA, which consists of the presence of antibodies in a serum sample and that is detected by competition with a specific antibody (mono- or

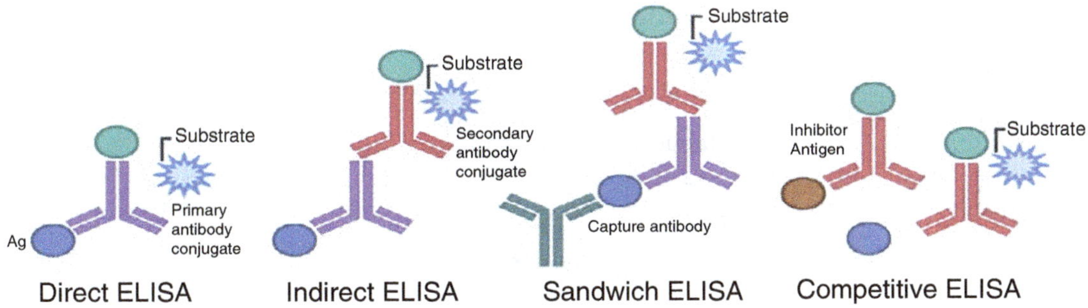

Fig. 1 Basic principles of the ELISA technique, including the main types of ELISA

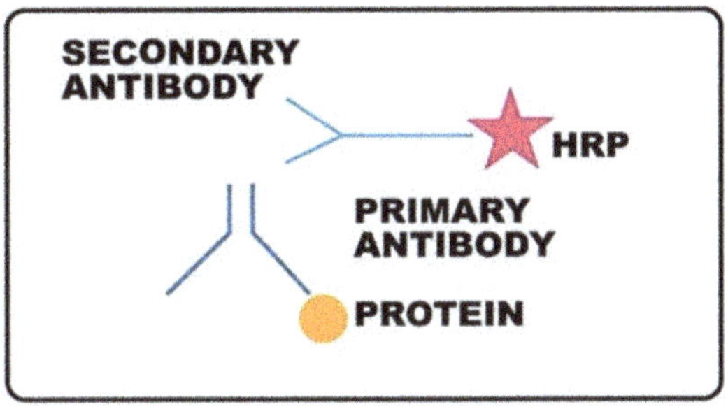

Fig. 2 Peroxidase technique applied to culture

polyclonal) directed against an antigen. Likewise, it is detected by the addition of a conjugate, but the color will appear in the wells where there were no antibodies. Subsequently, enzymes are added, which change the color of the medium by the reaction, and the amount is quantified by optical density.

The latter is used less, but is no less important as it is the method of choice when the molecule of interest has a few or very small binding epitopes.

Enzymes can be of several types; if it is peroxidase, this complex is usually peroxidase–antiperoxidase (*see* Fig. 2).

2 Materials

Incubator, spectrophotometer, microplate reader, vortex stirrers, micropipettes, tips, test tubes, distilled water, stopwatch, and absorbent paper.

Decidual stromal cell (DSC) supernatants (*see* **Notes 1** and **2**).

2.1 Determination of IL-6 by ELISA

1. IL-6 capture antibody.
2. ELISA microplate (Corning Costar 9018).
3. Coating buffer.

4. IL-6 protein standards.
5. Sample diluent.
6. Wash buffer.
7. Detection antibody (biotinylated anti-IL-6).
8. Avidin-HRP.
9. TMB.
10. Stop solution (2 N, H_2SO_4).

2.2 Determination of VEGF-R1 (sFlt-1) by ELISA

1. VEGF-Rl ELISA kit.
2. Wash buffer.
3. Protein concentrator.
4. Detection antibody (streptavidin-HRP anti-VEGF R1).
5. TMB.
6. Stop solution (1 M, H_3PO_4).

3 Methods

3.1 Determination of IL-6 by ELISA

A commercial ELISA kit was used for the detection of IL-6. The IL-6 concentrations in DSC/ESC cell supernatants were measured with the human IL-6 ELISA Ready-SET-Go! kit, reference: 88-7066 (Affymetrix, CA, USA). The sensitivity of the kit allowed the detection of IL-6 in a range between 2 and 200 pg/mL. Supernatants were collected and concentrated 10× in concentrators purchased from Minicon-Merck. The test was performed using the ELISpot method. ELISpot 5× is a diluent that binds to the non-binding site of the antibody (blocker). This allows the standard and sample to specifically recognize the antibody and reduces nonspecific binding of similar components (*see* **Note 3**).

1. Dilute IL-6 capture antibody in 1× coating buffer.
2. Deliver 100 μL/well of the capture antibody to ELISA. Incubate overnight at 4 °C.
3. After time had elapsed, aspirate wells and wash three times with 250 μL/well wash buffer. Soak for 1 min in each washing step to eliminate the loosely bound antibodies not adhering to the plate.
4. Drain the plate on absorbent paper to remove any residual buffer.
5. Dilute ELISA/ELISpot reagent (5×) with distilled water to prepare 1× diluent.
6. Deliver 200 μL/well and incubate 1 hour at room temperature.
7. Aspirate and wash once more with the washing buffer.

8. Reconstitute lyophilized standards with distilled water and allow to stand for 15 min with gentle agitation.
9. Prepare serial dilutions of the reconstituted standards in 1× diluent to establish a standard curve with a total of eight points in concentration.
10. Deliver 100 μL/well in duplicate of each concentration. Prepare two wells with 100 μL/well of 1× diluent as blanks.
11. Prepare samples of the DSC supernatants in diluent.
12. Add 50 μL of assay buffer to those wells designated for samples, and then add 50 μL of sample to each well in duplicate.
13. Seal the plate and incubate overnight at 4 °C.
14. Afterward, remove the samples from the wells. Wash well as in **step 3**, repeating for a total of five washes.
15. Add 100 μL/well of detection antibody prepared in 1× diluent. Place the plate in the dark or cover. Incubate at room temperature for 1 hour.
16. Antibody wells and wash as in **step 3**, repeating for a total of five washes.
17. Deliver 100 μL/well of the Avidin-HRP conjugate in 1× diluent. Seal the plate and incubate in the dark at room temperature for 30 min.
18. Aspirate and wash as in **step 3**, except that in this step the wells are to be soaked in the wash buffer for 2 min before aspiration. Repeat this step for a total of seven washes.
19. Add 100 μL/well of 1× TMB solution. Incubate the plate at room temperature for 15 min.
20. Deliver 50 μL of stop solution (sulfuric acid—2N H_2SO_4) to each well.
21. Read the plate at 450 nm using a colorimetric plate reader.
22. For data analysis, subtract the value of the blank and interpolate the absorbance data on the standard curve for the determination of concentrations.
23. Samples are multiplied by the dilution factor of 2 to calculate initial concentration.

3.2 Determination of VEGF-R1 (sFlt-1) by ELISA

A commercial ELISA kit was used for the detection of VEGF-R1 (sFlt-1). VEGF-R1 (sFlt-1) concentrations in the supernatants of DSC/ESC cells were measured with the human VEGF-R1 (sFlt-1) ELISA Platinum kit, reference: BMS268/3 (Affymetrix, Vienna, Austria). The sensitivity of the kit allowed detection of VEGF-R1 in a range starting at 0.03 pg/mL. Supernatants were collected and concentrated at 10× in concentrators purchased from Minicon-Merck trading house. The assay was performed following the manufacturer's instructions as shown below (*see* **Note 3**).

1. Coat a microplate with VEGF-R1 capture antibody supplied in the kit as previously described following the manufacturer's directions.
2. Wash plate wells twice with wash buffer.
3. Prepare standards starting at 40 ng/mL in assay buffer (1×) and serial dilute to achieve a seven-point dilution curve. Deliver 100 μL/well in duplicate for each concentration. Add 100 μL of assay buffer to duplicate blank wells.
4. Prepare DSC supernatant samples in diluent.
5. Add 50 μL of assay buffer to those wells designated for samples, and then add 50 μL of sample to each well in duplicate.
6. Deliver 50 μL/well of biotin-conjugated anti-VEGF-R1.
7. Incubate for 2 h at room temperature (25 °C).
8. Prepare antibody conjugated with streptavidin-HRP for use.
9. Wash wells six times with the wash buffer and discard the biotin-conjugated anti-VEGF-R1 antibody that remained free or did not bind.
10. Add 100 μL of streptavidin-HRP-conjugated antibody to all wells. Cover the plate incubated 1 h at room temperature 25 ° C.
11. Wash wells six times with wash buffer removing the streptavidin-HRP-conjugated antibodies that did not bind.
12. Deliver 100 μL of TMB substrate to all wells. Incubate for 30 min at room temperature (25 °C).
13. Add 100 μL of stop solution (phosphoric acid—1M H_3PO_4) to terminate the reaction.
14. Read the plate at 450 nm using a colorimetric plate reader.
15. For data analysis, subtract the value of the blank and interpolate the absorbance data on the standard curve for the determination of concentrations of samples.
16. Samples are multiplied by the dilution factor of 2 to calculate initial concentration.

4 Notes

1. *Subject groups*

   Pregnant women diagnosed with preeclampsia (blood pressure > 140/90 mmHg and proteinuria >300 mg/L in 24-h urine) were monitored in the Obstetrics Unit at the ESE-University Hospital of Santander and the Chicamocha clinic (PE group, $n = 19$). As control, we included two groups: (1) uncomplicated vaginal deliveries (VD group, $n = 16$)

managed at the Hospital del Norte, the Santa Teresita Mother-Child Immediate Unit (UIMIST) and the Chicamocha SAS of the city of Bucaramanga, Santander (Colombia), and (2) pregnant women who required elective cesarean section for the labor (CE group, $n = 13$). In all groups, pregnant women who had vaginal infections were excluded from the study. All subjects completed informed consent forms for collection of tissue samples, which were collected according to the requirements of the Universidad de Santander ethics committee in Bucaramanga, Colombia.

2. *Cell isolation and culture*

   Basal decidua tissue was separated from the placentas upon delivery. To establish DSC lines, tissues were minced, digested with 5 mg/mL collagenase IA (Sigma-Aldrich) for 15 min at 37 °C, and centrifuged at 450×*g* for 10 min. The cell pellet was resuspended in PBS and centrifuged on Ficoll-Paque (Sigma-Aldrich) for 20 min at 600×*g*. Decidual cells were collected from the interphase, suspended in PBS, and washed. The resulting suspension was incubated in culture flasks for 24 h at 37 °C with 5% CO_2 in Opti-MEM (Invitrogen) supplemented with 3% fetal calf serum (FCS) and antibiotics. After overnight incubation, nonadherent cells in the supernatant were discarded. After 1–3 weeks, adherent cells were morphologically uniform and covered the whole surface of the culture flask. The purity of DSCs was confirmed by flow cytometry. Conditioned supernatants were collected from the DSC cultures at day 23 with 80% confluence.

   Supernatants were collected from DSC cultures at 80% confluence into sterile 15-mL falcon tubes, then labeled and stored at –20 °C for subsequent cytokine analysis.

3. *General precautions and recommendations*

   (a) At the moment the substrate is added, immediately the wells should be capped from the light.

   (b) The wells should be well washed to avoid contamination with reagents.

   (c) A good homogenization of the samples with the conjugate should be performed.

   (d) It is not recommended to use tape on the lower part of the wells to fix the plates due to the residue of glue that they leave.

   (e) Change the pipette tip between dispensing different samples/reagents to avoid contamination of the reagents used.

   (f) Ensure that the filling of the wells at the time of washing is equal for all wells that have samples, be careful not to generate bubbles during the dispensing of the washing solution, and do not overflow the wells until they come out of their overflow, as well as the correct drying of the wells.

(g) If bubbles formed, use a pin or needle to pop the bubbles.

(h) Before using the substrate confirm that it is not contaminated. Color change indicates contamination.

(i) The incubation time of the substrate with the sample is the key to the detection limit and should not exceed the time specified by the protocol.

(j) Before dispensing the conjugates, they should be at room temperature.

(k) When using the stop solution, do not overshoot the reading after 30 min because false results may occur.

(l) If possible, the technique should be run at room temperature.

(m) During the incubation period, the wells should be covered to avoid any contamination or evaporation.

References

1. Konstantinou GN (2017) Enzyme-linked immunosorbent assay (ELISA). Methods Mol Biol 1592:79–94. https://doi.org/10.1007/978-1-4939-6925-8_7
2. Aydin S (2015) A short history, principles, and types of ELISA, and our laboratory experience with peptide/protein analyses using ELISA. Peptides 72:4–15. https://doi.org/10.1016/j.peptides.2015.04.012. Epub 2015 Apr 20
3. Tighe PJ, Ryder RR, Todd I, Fairclough LC (2015) ELISA in the multiplex era: potentials and pitfalls. Proteomics Clin Appl 9(3–4):406–422. https://doi.org/10.1002/prca.201400130. Epub 2015 Mar 25. PMID: 25644123; PMCID: PMC6680274
4. Engvall E, Perlmann P (1971) Enzyme-linked immunosorbent assay (ELISA). Quantitative assay of immunoglobulin G. Immunochemistry 8(9):871–874. https://doi.org/10.1016/0019-2791(71)90454-x
5. Ong DSY, Fragkou PC, Schweitzer VA et al (2021) How to interpret and use COVID-19 serology and immunology tests. Clin Microbiol Infect 27(7):981–986. https://doi.org/10.1016/j.cmi.2021.05.001. Epub 2021 May 8. PMID: 33975005; PMCID: PMC8106522
6. Beavis KG, Matushek SM, Abeleda APF et al (2020) Evaluation of the EUROIMMUN anti-SARS-CoV-2 ELISA assay for detection of IgA and IgG antibodies. J Clin Virol 129:104468. https://doi.org/10.1016/j.jcv.2020.104468. Epub 2020 May 23. PMID: 32485620; PMCID: PMC7255182
7. Kohl TO, Ascoli CA (2017) Direct competitive enzyme-linked immunosorbent assay (ELISA). Cold Spring Harb Protoc (7):pdb.prot093740. https://doi.org/10.1101/pdb.prot093740
8. Lin AV (2015) Indirect ELISA. Methods Mol Biol 1318:51–59. https://doi.org/10.1007/978-1-4939-2742-5_5
9. Hornbeck PV (2015) Enzyme-linked immunosorbent assays. Curr Protoc Immunol 110:2.1.1–2.1.23. https://doi.org/10.1002/0471142735.im0201s110

Chapter 9

Gyrolab Immunoassays: Miniaturization, Automation, and Integration into a Rapid Workflow

Johan Engström

Abstract

Gyrolab® is an open immunoassay platform that automates the complete immunoassay protocol in a microfluidic disc. The column profiles generated with Gyrolab immunoassays are used to gain more information about biomolecular interactions that can be useful in assay development or quantify analytes in samples. Gyrolab immunoassays can be used to cover a broad concentration range and diversity of matrices in applications ranging from biomarker monitoring, pharmacodynamics and pharmacokinetics studies, to bioprocess development in many areas, including therapeutic antibodies, vaccines, and cell and gene therapy.

This chapter is an overview of Gyrolab technology, including system components and the assay development workflow, including the process of selecting affinity reagents, Gyrolab Bioaffy CDs, and assay conditions to optimize immunoassays. Two case studies are included. The first involves an assay for the humanized antibody pembrolizumab used in cancer immunotherapy that can generate data for pharmacokinetics studies. The second case study involves quantification of the biomarker and biotherapeutic interleukin-2 (IL-2) in human serum and buffer. IL-2 has been implicated in the cytokine storm associated with COVID-19, and cytokine release syndrome (CRS), which can occur during chimeric antigen receptor T cell (CART) therapy used in treating cancer. These molecules also have therapeutic relevance in combination.

Key words Immunoassay, Biotherapeutics, Therapeutic antibody, Pharmacokinetics, Pharmacodynamics, Biomarker, Cytokines, Bioprocess development, Affinity screening, High-throughput immunoassays, Cell and gene therapy, Ligand-binding assays

1 Introduction

Immunoassays are widely utilized to quantify antigens in many fields, including pharmaceutical and vaccine development, cell and gene therapy, and diagnostics. Engvall and Perlmann first described the ELISA methodology 50 years ago, and immunoassay technology has continuously improved ever since [1]. This has led to the development of automated systems and single-use devices for rapid and convenient analysis in many applications. Gyrolab

Robert S. Matson (ed.), *ELISA: Methods and Protocols*, Methods in Molecular Biology, vol. 2612,
https://doi.org/10.1007/978-1-0716-2903-1_9,

technology has been developed with the aim of speeding up analysis, reducing the consumption of samples and reagents, and automating liquid handling. The inherent flexibility of ELISA has been maintained, including all assay steps, and sample preparation steps can be integrated. ELISA immunoassays are often readily transferred to Gyrolab systems using the same reagents, with the Gyrolab assay frequently providing improved dynamic range and lower variability.

2 Overview of Gyrolab Technology

Gyrolab immunoassays are based on a nanoliter-scale flow-through technology that enables rapid analysis of samples in parallel, with a time-to-result of approximately 1 h for ~100 samples. With the flow-through format, the contact time between the sample and the solid phase is short, which minimizes matrix interference. The platform is based on a compact disc with preloaded streptavidin columns used for the affinity interaction ([2], *see* Fig. 1). The instrument and integration of operations in the microfluidic device together reduce the hands-on time for the preparation of samples and reagents. While kits and software are available for specific applications and assays, the Gyrolab platform is an open system that supports the flexible use of immunoassay reagents for any application. Assay optimization is rapid, based on a range of assay formats and optimized Gyrolab methods that include all steps in the workflow for liquid handling and detection.

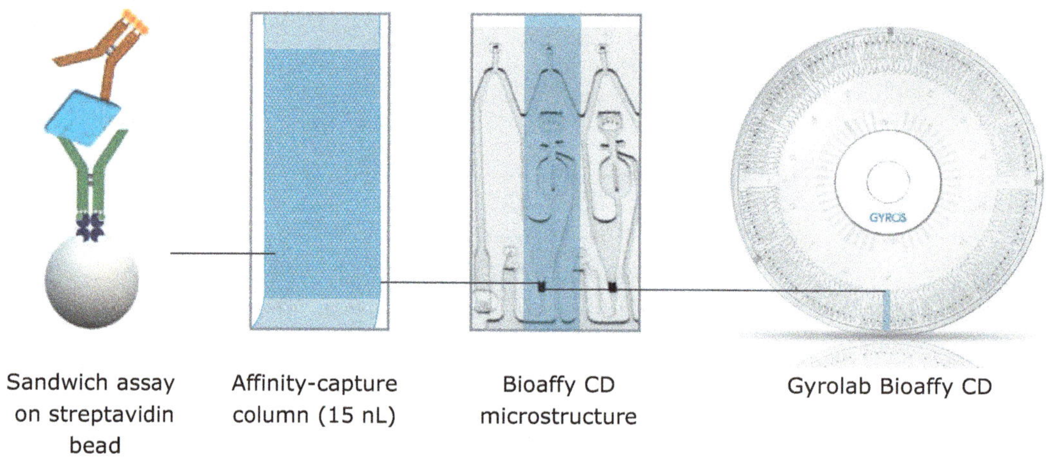

Fig. 1 Gyrolab Bioaffy CDs include all essential functions performed in immunoassays in a miniaturized format. The nanoliter-scale flow-through technology uses capillary and centrifugal forces to control liquid flow through an affinity column comprising streptavidin-coated beads or sample-reagent mixing in a specially designed CD

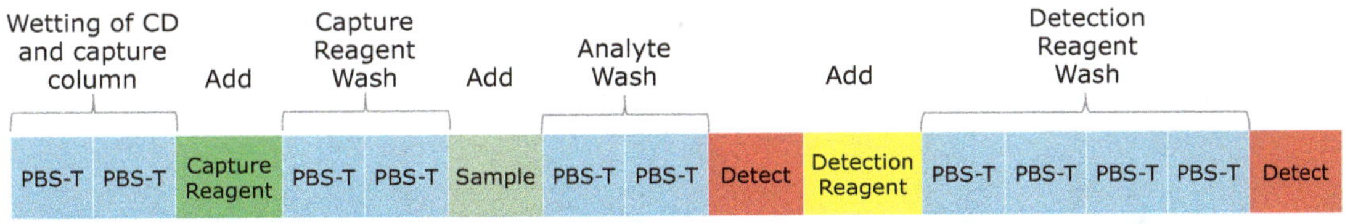

Fig. 2 Assay steps described in the Gyrolab methods are performed serially and automate the liquid handling robot and microfluidic operations. This example is a three-step immunoassay method that loads capture reagent, sample, and detection reagent followed by laser-induced fluorescence detection

The quantitative data obtained from Gyrolab assays is equivalent to traditional ELISA. Gyrolab technology also provides fluorescent column profiles that give more information about antibody affinity and a quality assessment of the result. This unique functionality is valuable when screening for antibody reagents and improving immunoassay robustness.

Gyrolab methods are instrument instructions in the Gyrolab software that replace the manual assay protocols to automate the workflow in a flexible way (*see* Fig. 2). The automation provided by predefined methods and Gyrolab software ensures smooth immunoassay set up, day-to-day reproducibility, rapid evaluation of the results, and smooth transfer of assays between labs with maintained performance. Predetermined alternative methods, including washing procedures or methods to vary sample and reagents flow speed over the column, are available to enable the development of a variety of assay formats (including one-step and two-step assays) to solve challenging analytical situations.

Gyrolab immunoassay platform can be used together with ready-to-use kits or assays can be designed using in-house affinity reagents. In addition to the affinity reagents, materials such as Gyrolab Bioaffy CDs, dilution buffers, and system liquids are required.

2.1 Gyrolab Bioaffy CDs

Gyrolab Bioaffy CDs are ready to use without any need for preparation. The microfluidic CDs are available in several versions suitable for different applications and analytical requirements. These CDs differ primarily in sample volume, which usually determines the sensitivity of the assay (*see* Fig. 3). Sample volumes are small compared to conventional immunoassays and range from 20 to 4000 nL. An additional variable is the capacity of the affinity column, which can include either standard capacity or high-capacity (HC) porous beads. HC columns are useful for high analyte concentrations (Gyrolab Bioaffy 20HC) or when polyclonal or low-affinity capture reagents are used (Gyrolab Bioaffy 1000HC).

For certain applications, the Gyrolab Mixing CD is useful for sample pretreatment prior to solid-phase interaction with the affinity column. This can include acid treatment or enzymatic steps that are integrated into the assay protocol with a Gyrolab method.

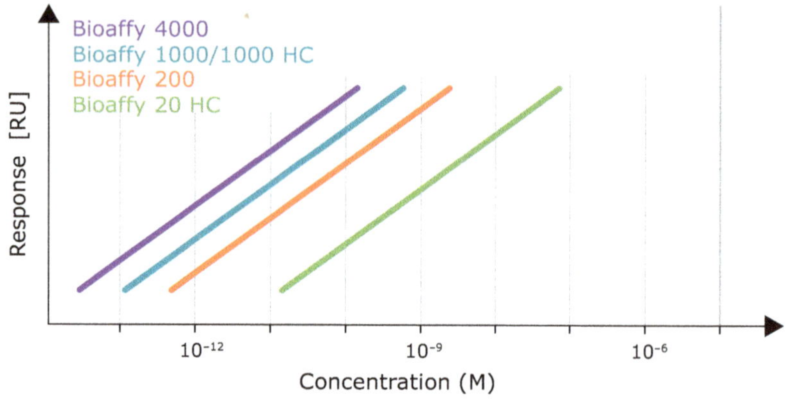

Fig. 3 Gyrolab Bioaffy CDs cover sample volumes ranging from 20 nL to 4000 nL that provide a broad dynamic range for assays where high-sensitivity or high-concentration measurements are needed

Fig. 4 Gyrolab xPlore and Gyrolab xPand

2.2 Gyrolab Systems

Gyrolab systems are liquid-handling robots that transfer sample and reagents from microplates to the microfluidic CD, where the immunoassay is performed. About 100 samples are analyzed in parallel depending on the CD type and using either the single CD system, Gyrolab xPlore, or the larger system Gyrolab xPand, which enables automated sequential handling of five CDs for larger number of samples, including cooling to ensure sample and reagent stability (*see* Fig. 4).

2.3 Gyrolab System Buffers

Gyrolab systems need Pump Liquid as a hydraulic liquid and wash solution that is used in the wash station to clean the liquid transfer needles.

Gyrolab Bioaffy Pump Liquid and Gyrolab Wash Station Solution 1 contains 15 mM phosphate buffer, 150 mM NaCl, 0.05% Proclin 950 with 0.01% Tween pH 7.4.

Different Gyrolab Bioaffy Wash Station Solutions can be used, but the high-performance Gyrolab Bioaffy Wash Buffer pH 11 can be used for a wide range of applications, including with challenging analytes that have a tendency for carry over in the liquid transfer needles.

2.4 Assay Buffers

The sample dilution buffer can have a major effect on assay performance and screening to find the optimal buffer is recommended. A wide range of Rexxip® buffers have been developed to meet the needs of different applications or match the physicochemical properties of the analyte or the sample. These buffers are optimized for the microfluidic immunoassay and contain different additives, such as detergents for hydrophobic analytes, salt for highly charged analytes, and blockers to reduce heterophilic interactions.

2.5 96-Well Plate

Samples and reagents are loaded in a 0.2-mL skirted PCR plate with a microtiter plate foil to reduce evaporation. This plate is loaded in the Gyrolab system.

2.6 Gyrolab Assay Development Workflow

Gyrolab assay development is efficient due to the short turnaround time and the predefined Gyrolab methods that simplify the optimization process. Optimization of an immunoassay includes the selection of an assay format and Gyrolab method. The most important parameters in assay development are the screening of affinity reagents and the combination of capture and detection reagents.

2.6.1 Selection of Gyrolab Method

Parameters for transfers of liquids (samples, reagents, washes) and spin sequences are preprogrammed in the Gyrolab method. Different methods are selected depending on the CD and assay design. The most widely used protocol is the generic three-step assay that involves sequential addition of reagents, samples, and wash buffers, and then performing the detection operations (*see* Fig. 2). The three-step method is useful for a wide range of applications using different reagents (*see* Fig. 5). The two case studies, used for pharmacokinetics studies and biomarker analysis, presented below are based on the three-step method.

Several assay formats are available for the pharmacokinetic analysis of a therapeutic monoclonal antibody in human matrices. In preclinical assays for animal samples, the indirect antibody format with generic anti-human antibody for detection can be used. For clinical assays in human matrix, the bridging assay format using the target molecules or an anti-idiotypic antibody as both capture and detection reagent are preferred.

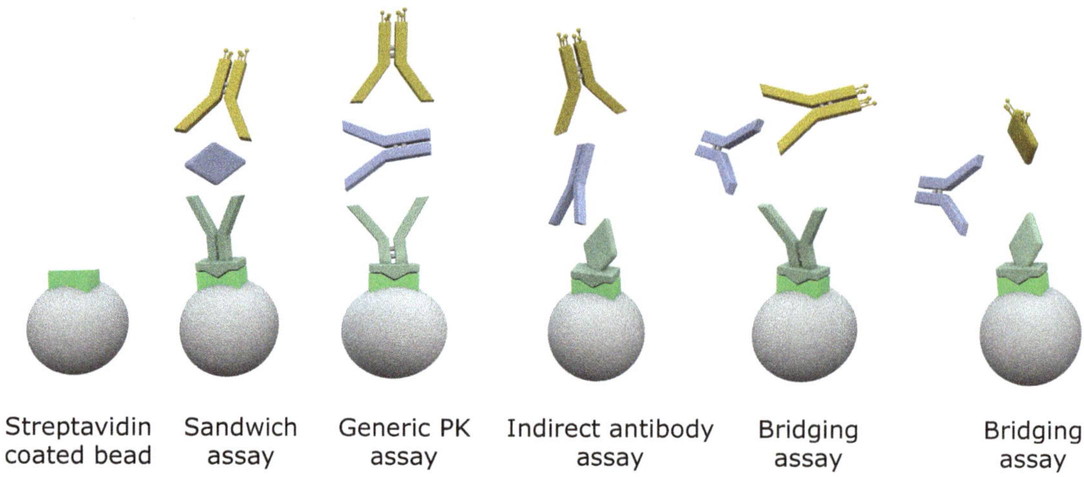

Fig. 5 Various assay formats can be readily performed using Gyrolab three-step methods. The generic streptavidin-coated bead can be functionalized with any biotinylated reagent and used in combination with a suitable detection reagent

Table 1
Gyrolab Bioaffy CDs and methods using three-step immunoassays with two wash solutions

CD	Sample volume (nL)	Method name
Gyrolab Bioaffy 20 HC	20	20HC-3W-011-A
Gyrolab Bioaffy 200	200	200-3W-002-A
Gyrolab Bioaffy 1000	1000	1000-3W-006-A
Gyrolab Bioaffy 1000 HC	1000	1000HC-3W-011-A
Gyrolab Bioaffy 4000	4000	4000-3W-001-A

In the biomarker assay, the same Gyrolab method can be used but the reagents are based on specific antibodies for the selected biomarker molecule—see the sandwich assays in Fig. 5.

2.6.2 Design Run

Gyrolab Control or Manager software is used to create a new Run. In this process, all the information to perform a Gyrolab immunoassay is defined. This will generate an executable Run and a Loading List with the layout of samples, reagents, and wash buffers on the microplate.

The appropriate three-step method is selected in the Gyrolab method database for the CD to be used in the assay. Three-step methods using two wash solutions for each Gyrolab Bioaffy CD are listed in Table 1.

The software automatically guides the process of defining the capture and detection reagents, standard series, and control samples. Each reagent is defined and capture and detection reagents can be combined to find optimal reagents and concentrations for the assay. Standard samples are defined with concentrations specified based on suitable dilution steps. Different experiments using different dilutions, buffers, analytes, etc., can be designed.

2.6.3 Capture and Detection Reagents

Capture and detection reagents can be antibodies, other proteins, peptides, or nucleic acid-based binders, either commercially available or developed in-house. The capture reagents must be labeled with biotin, and the detection reagent must be labeled with a fluorescence tag suitable for detection with a 635-nm red laser, e.g., Alexa Fluor® 647. Both of these conjugations can be performed using a wide range of commercial kits. Antibodies can be biotinylated or fluorescently labeled with *N*-hydroxy succinimidyl (NHS) chemistry for conjugation to free amine groups.

The labeling procedure should contain a separation step that removes free coupling reagent after conjugation. Free biotin will reduce capacity on the solid phase and free fluorophore will increase assay background. When using NHS chemistry, the reagents must be free from amine compounds, including stabilizing proteins, for the best preparation of conjugates.

The recommended labeling procedures are as follows:

- Antibodies are biotinylated with EZ-Link Sulfo-NHS-LC-Biotin and separated with Protein Desalting Spin Column from Thermo Fisher Scientific.
- Antibodies are fluorescence labeled with Alexa Fluor 647 Labeling or DyLight® 649 Labeling Kit from Thermo Fisher Scientific.

Identification of Antibody Reagents

The assay performance for any immunoassay is greatly influenced by the characteristics of the reagents available for the assay. There are several properties to consider when selecting affinity reagents, including affinity, specificity, cross-reactivity, and purity. Methods generating monoclonal antibodies are continuously improving and clones with high-affinity binding are usually commercially available. For polyclonal reagents, antigen affinity purification is recommended. Reagents used in an existing plate-based assay are often a good starting point. Orientation of capture or detection antibodies or the antibodies themselves may need to be changed for optimal performance in Gyrolab assays.

Screening Affinity Reagents

There are different ways to screen antibodies with Gyrolab system. For optimization of sandwich immunoassays, it is recommended that each affinity reagent is prepared with both biotin and fluorescence label and the optimal binding pair is selected for quantification analysis. In early-stage clone screening, a large number of

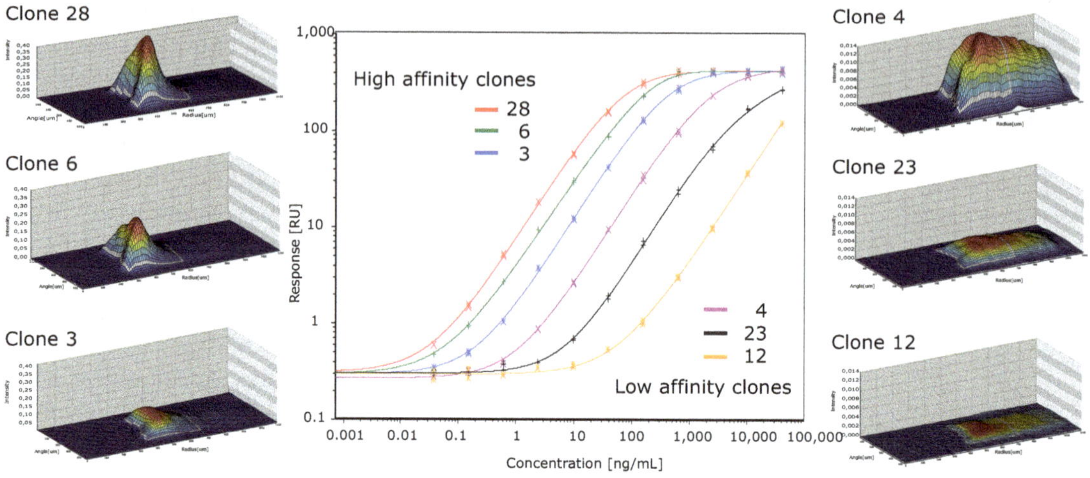

Fig. 6 Column-binding profiles are valuable for screening antibody clones to select candidates for the development of robust assays. The column profiles represent the signal intensity from the fluorescent immune complex captured on each microfluidic column. The column profile has a characteristic shape, with enrichment of the immune complex at the beginning of the column. High-affinity clones have a dose–response that is shifted to the left and sharp peaks. Curves to the right with broad peaks represent lower affinity interactions

clones can be evaluated rapidly if the biotinylated target is on the column. Using Gyrolab Bioaffy 200, up to seven antibody pairs can be screened using a seven-point standard curve in singlicate (plus blank). Up to 14 antibody pairs can be screened using a four-point standard curve in duplicate.

Ranking Clones Using Standard Curves and Column Profiles

Standard curves of different candidates can be compared regarding signal-to-noise level and curve shape at high and low concentrations. The column-binding profiles give more information about the interaction, with high-affinity interactions generating sharp peaks (Fig. 6). The sample is efficiently enriched in the column, and this is the optimal condition for obtaining a sensitive immunoassay. For lower affinity interactions, a broad peak is generated and not all analyte molecules will be captured in the column. This assay will have lower sensitivity that can be improved using reagents with higher affinity.

Anti-idiotypic antibodies bind to the idiotype located in the variable region of another antibody. These anti-idiotypic antibodies are widely used to study therapeutic monoclonal antibodies as reagents in pharmacokinetics assays and as positive controls in immunogenicity characterization. Figure 6 illustrates the binding properties of six antibodies. The column-binding profiles generated correlate with the information from dose–response curves for affinity ranking and characterization.

If high-affinity reagents are not available, one option is to enrich the sample at the top of the column by using the porous HC particle with a higher capture reagent density (the "HC" Bioaffy CD series—*see* below). Another method is to adjust the liquid flow speed of the sample and detection reagent, which increases the contact time for the solid-phase interaction. If the time is still too short to create the capture–analyte molecular complex, a two-step assay can be useful in which the capture–analyte interaction is performed offline in a microplate before using the high-affinity streptavidin–biotin interaction in the microfluidic CD.

2.6.4 Improved Assay Performance with Bioaffy 1000HC

For antibodies with lower affinities, the column profiles become sharper using a CD with the porous HC particle. For example, Fig. 7 shows column profiles for a rat erythropoietin (EPO) assay using both Gyrolab Bioaffy 1000 and Gyrolab Bioaffy 1000HC for both recombinant target and endogenous target. The profiles indicate the improved enrichment using the HC particle and the reduced reagent affinity for the endogenous target.

Column profiles are very useful for quality assessment when quantifying unknown samples. Matrix effects or other issues can be addressed to avoid erroneous interpretation of results.

2.6.5 Titration of Reagents

Titration of capture and detection reagent may influence the assay performance or can be an additional tool to further reduce the consumption of the affinity reagents.

Capture Reagent

Saturating the streptavidin affinity column requires about 200 nl of 100 μg/mL (~700 nM) of biotinylated capture reagent. For low-molecular-weight capture molecules, the concentration should be increased to obtain full saturation. For bridging assays (see Fig. 5), saturation may cause both arms of a drug antibody analyte to bind to the column, making it inaccessible for detection. In these cases, binding density can be adjusted by replacing part of the biotinylated reagent with biotinylated BSA (*see* Fig. 8). For some assays, it is possible to titrate the capture reagent down to 5% of a saturated column with maintained performance. This may be attractive for precious reagents.

Detection Reagent

The detection reagent is labeled with Alexa Fluor 647 and diluted in Rexxip F, usually at concentrations in the range of 5–25 nM. Before dilution of the detection reagent, the stock solution should be centrifuged at 12,000 × *g* for 4 min. Reducing the concentration reduces background but also the fluorescence response from the specific interaction. The signal-to-noise ratio at the lower part of the curve can be used as a selection criterion to identify the optimal concentration.

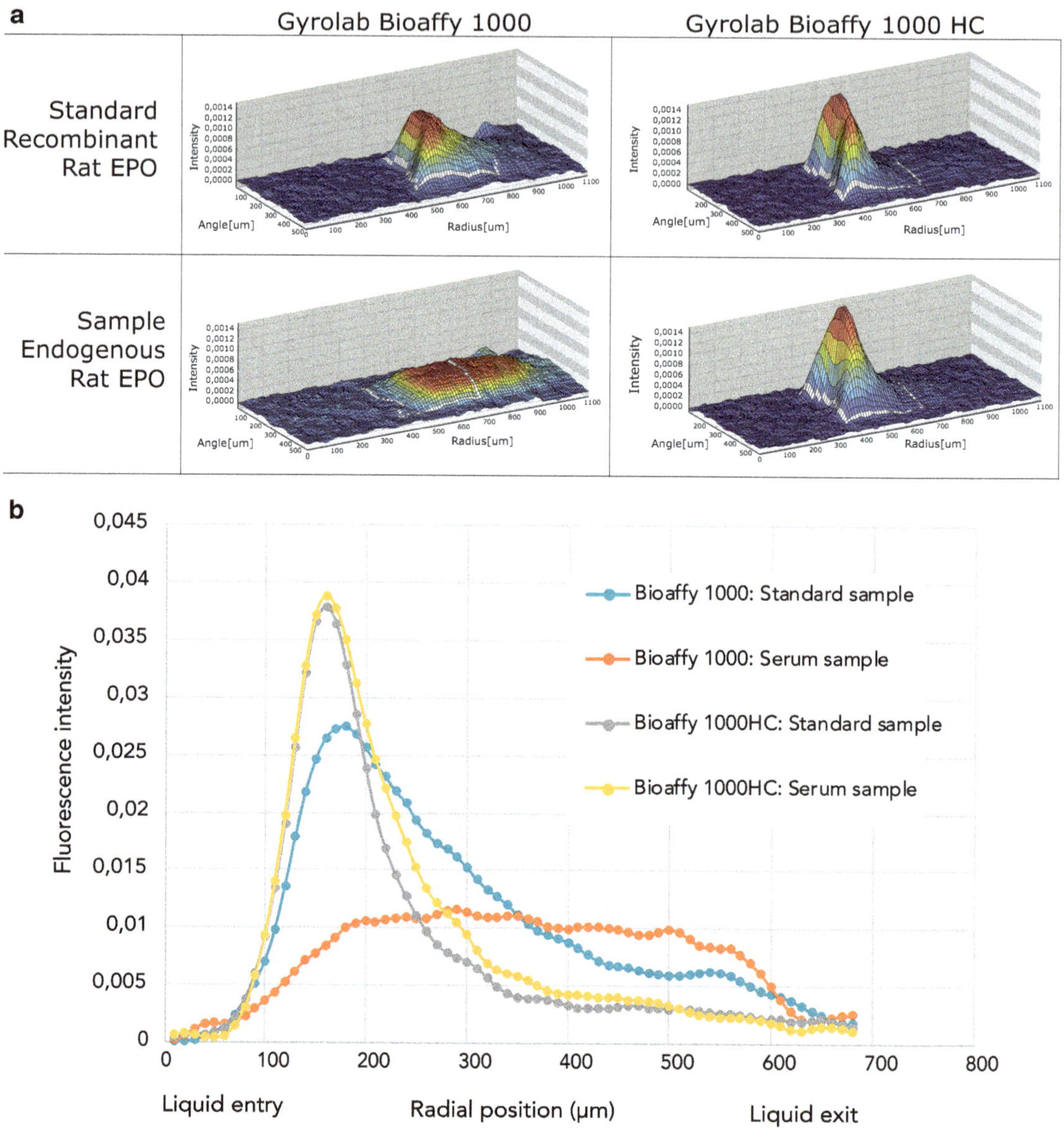

Fig. 7 (**a**) 3D column profiles from a rat EPO assay showing enriched analyte on the HC particle and reduced reagent affinity for the endogenous target. (**b**) 2D column profiles showing the distribution of fluorescence signal in the flow direction. Increasing the particle capacity minimizes the difference between standard sample with recombinant EPO and serum sample with endogenous EPO

2.6.6 Prepare Samples

Samples that might contain particulate material, such as cell debris or a pellet, are centrifuged 3000 × *g* for 15 min at 8 °C prior to diluting. If samples are frozen, they should be thawed and vortexed gently before centrifugation.

A suitable Rexxip buffer is selected depending on the sample type, species, and characteristics of the analyte.

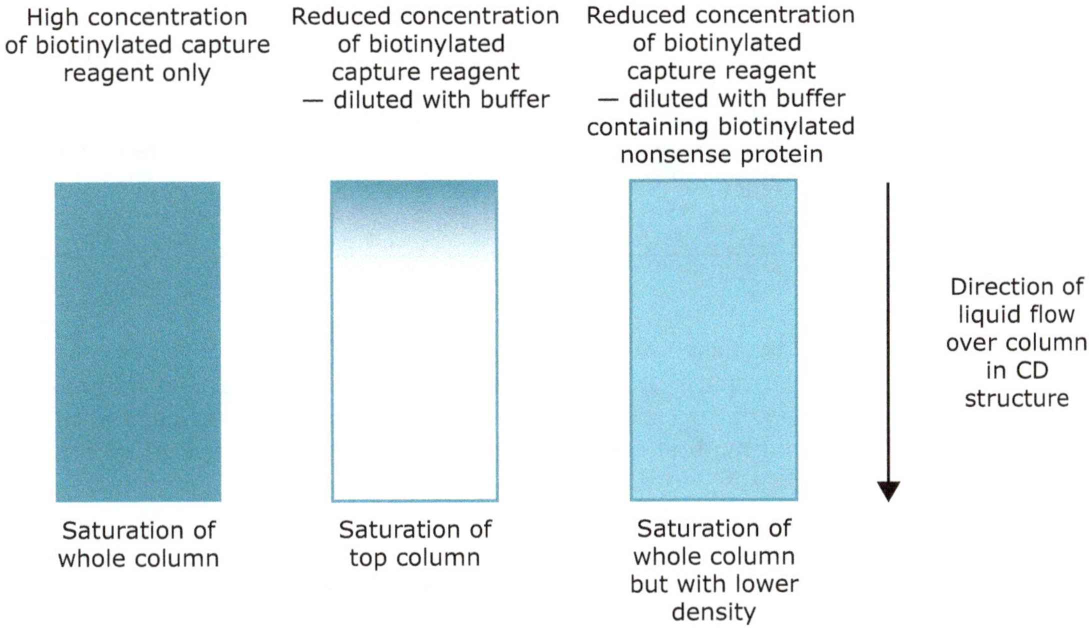

Fig. 8 The density of the capture target on the column can be regulated by loading with a combination of biotinylated molecules

2.6.7 Prepare a Run

A Gyrolab Run is created with Gyrolab Manager or Gyrolab Control software by combining a Gyrolab method with the experimental design that includes, for example, reagents, standard series, quality controls, and unknown samples.

Reagents, samples, and wash solutions are added to microplates according to a Loading List with information about microplate location and liquid volumes. Reagents include the capture and detection reagents, and PBS-T is used as wash solution. The microplate is covered with foil to reduce evaporation.

2.6.8 Data Analysis

Gyrolab Evaluator software facilitates the analysis of the results with a wide range of functions that included curve fitting, presentation of standard curves, and quantification of concentrations in samples. Performance characteristics, including recovery, limit of detection, and quantification for the assay, are presented by the software and in reports that are valuable when developing and validating assays. Analysis settings can be adapted according to analytical requirements. In the case studies presented below, default settings were used, including a five-parameter logistic model with the response as the weight.

The total integrated response of each column is used to quantify samples, but more information can be obtained. Gyrolab Viewer presents the fluorescence distribution in each affinity column to support the assessment of molecular interaction properties for affinity screening, quality control and also assay troubleshooting, e.g., to investigate outliers.

For some applications, data interpretation is more demanding and requires specialized software modules. The ADA software facilitates the immunoassay workflow used to measure levels of antidrug antibody (ADA) in the assessment of drug immunogenicity. The Affinity software is useful for the in-solution characterization of affinity and the determination of active concentrations.

3 Case Studies

The main aim of the PK assay for pembrolizumab was to measure the analyte at both low and high concentrations and minimizing matrix effects. A bridging assay was used to minimize matrix effects and method development involved adjustment of capture reagent density and detection reagent concentration to ensure high assay performance and robustness.

The focus of the biomarker/therapeutic IL-2 assay was to achieve high sensitivity for the analyte as a biomarker and shift the dynamic range depending on the development stage when quantifying the analyte as a therapeutic. Assay development was generally quite straightforward once the optimal reagents had been identified. The dynamic range could be shifted by simply changing the Bioaffy CD and associated method to meet specific analytical needs.

3.1 Materials

3.1.1 Pembrolizumab (Keytruda) PK Assay

Capture: 350 nM PD-1 (Abcam, Cat# ab174035), labeled with biotin + 350 nM biotinylated BSA in PBS-T.

Detection: Anti-pembrolizumab (Bio-Rad, Cat# HCA297) labeled with Alexa Fluor 647, 35 nM in Rexxip F.

Analyte: Keytruda (Merck Sharp & Dohme Corp) in Rexxip H with 25% human serum.

3.1.2 IL-2 Biomarker Assay

Capture: Antihuman IL-2 (BD Pharmingen, clone 5344.111, Cat# 555051) labeled with biotin, 100 μg/mL in PBS-T + 0.02% sodium azide.

Detection: Antihuman IL-2 (R&D Systems, clone 5334, Cat# MAB202–100) labeled with Alexa Fluor 647, 25 nM in Rexxip F.

Analyte: Recombinant human IL-2 in Rexxip H (R&D Systems, Cat# 202-IL-10).

3.2 Gyrolab Methods

The Gyrolab methods for each Bioaffy CD used in the case studies are shown in Table 1. These methods specify all the steps in the immunoassay workflow (*see* Fig. 2).

3.3 Pharmacokinetics Analysis: Pembrolizumab

3.3.1 Background

New therapeutics targeting immune checkpoint proteins have been significant for the treatment of cancer. One example is pembrolizumab (Keytruda®), which is a cancer immunotherapy biopharmaceutical that belongs to the group of programmed cell death protein 1 (PD-1) inhibitors. Pembrolizumab is a humanized antibody of IgG4 isotype that blocks a protective mechanism of cancer cells, allowing the immune system to destroy them [3].

The assay for pembrolizumab was set up as a bridging three-step sandwich assay with biotinylated human PD-1 as a capture molecule and a recombinant human anti-pembrolizumab labeled with Alexa Fluor 647 as a detection molecule.

3.3.2 Results of Method Development

Density of Capture Reagent

The density of the capture reagent on the solid phase is an important parameter to optimize for a bridging assay. In this assay, using only PD-1 as the capturing molecule increases the capacity but reduces sensitivity. The density can be reduced with biotinylated BSA to prevent both arms of the antibody analyte from binding to the immobilized target. Reducing the density makes the other arm accessible for detection and greatly improves the sensitivity (*see* Fig. 9). The combination finally selected was a 1:1 mixture of biotinylated PD-1 and BSA.

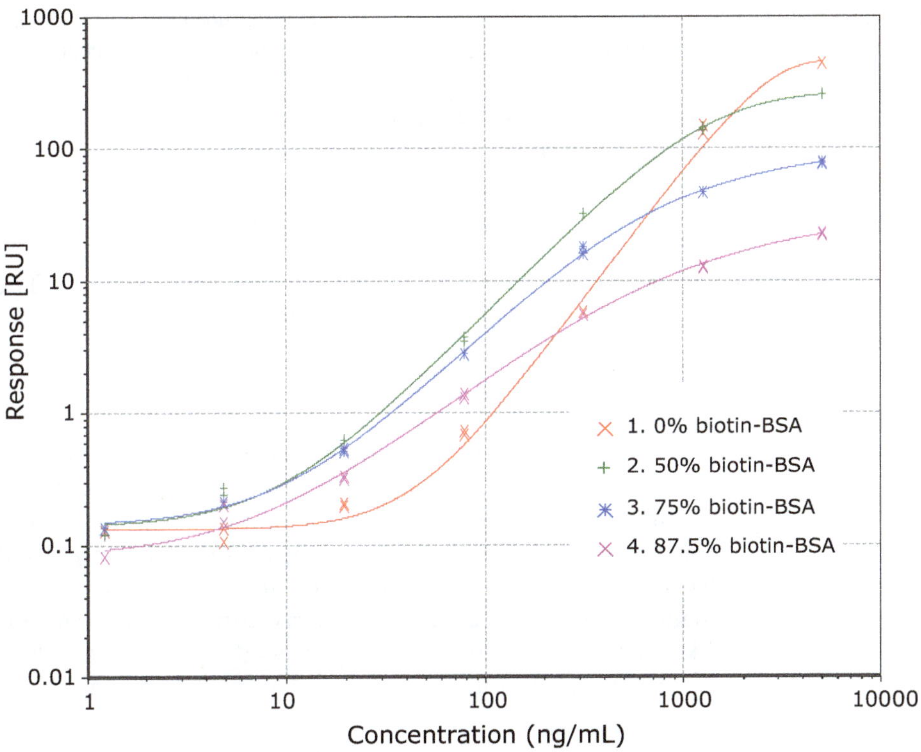

Fig. 9 Adjusting the surface density of the PD-1 capture reagent with biotinylated BSA improved the assay sensitivity

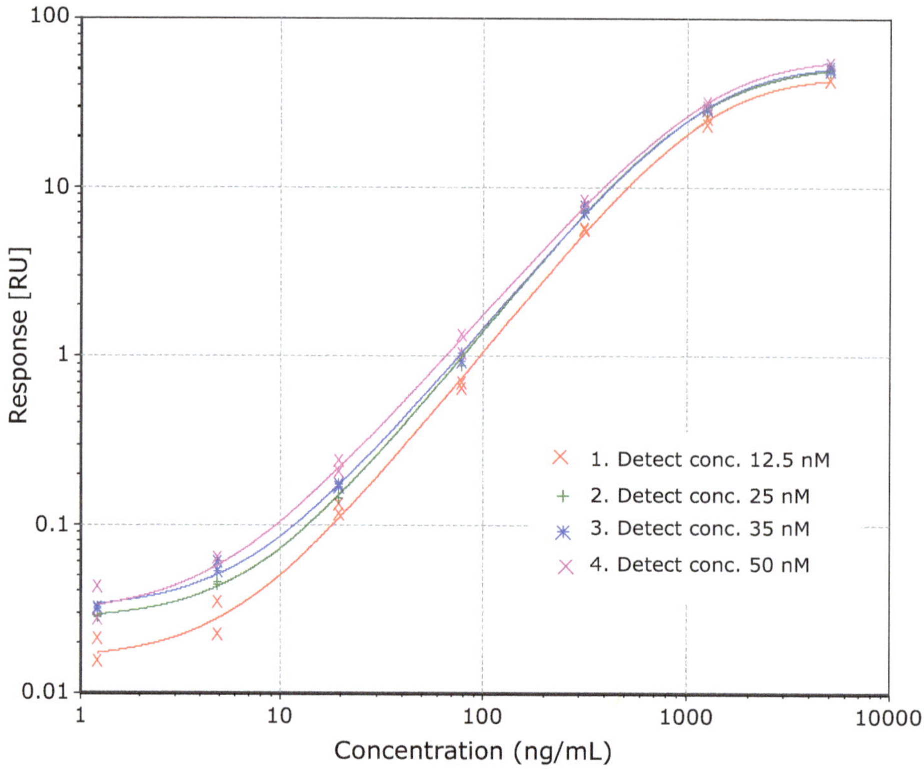

Fig. 10 Detection reagent was titrated in the range of 50–12.5 nM. The signal-to-noise ratio and the variation (CV) in the lower part of the curve are used as criteria for selecting the detection reagent concentration

Concentration of Detection Reagent

The concentration of the detection reagent can be fine-tuned to improve assay performance and robustness. The anti-idiotypic antibody detection reagent was titrated in the range of 50–12.5 nM. Lowering the concentration reduced the background but also increased the variation in the response (*see* Fig. 10). The most robust and sensitive assay was obtained using a 35-nM detection reagent.

Matrix Tolerance

Different assay formats are used to measure therapeutic antibodies. The bridging assay format is preferred for measurements in human serum using the antibody paratope (antigen binding site) for both the capture and detection interactions. This assay design is resistant to interference from bulk IgG in human serum and the performance is not affected by the matrix concentration (*see* Fig. 11). Matrix tolerance also reduces the need to dilute samples, which increases assay sensitivity and simplifies sample preparation.

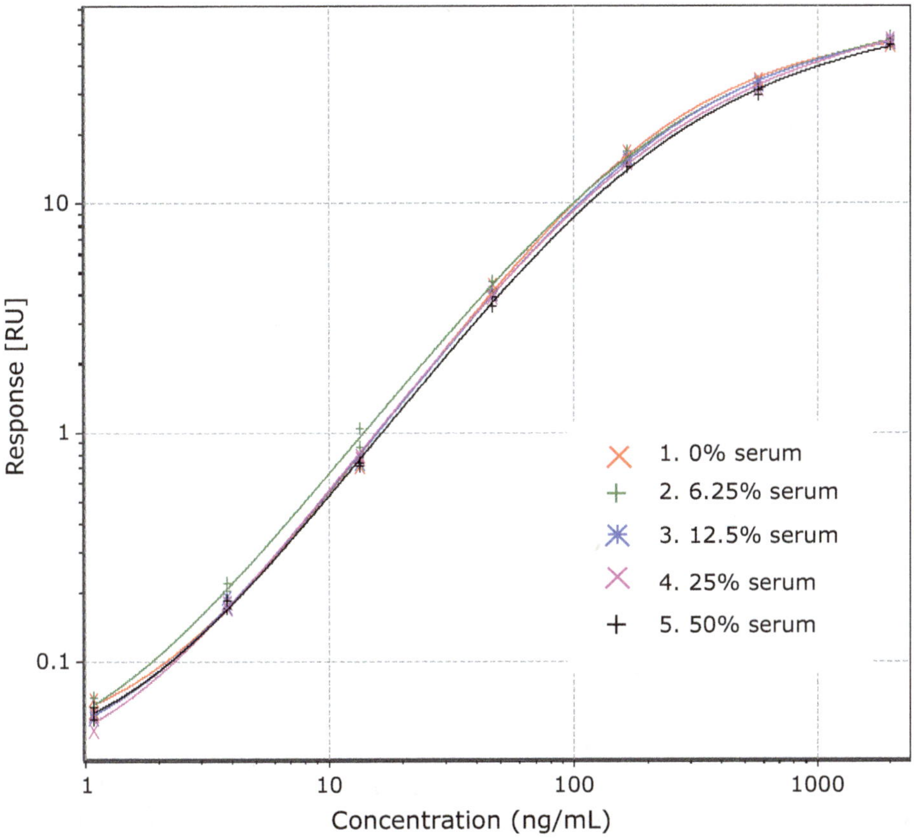

Fig. 11 Human serum spiked with pembrolizumab diluted with Rexxip H. No matrix effect was observed in this bridging assay

Choice of Gyrolab CD to Meet Analytical Needs

Using the assay protocol developed for pembrolizumab, different Gyrolab CDs depending on the analytical expectation of the assay can be used (*see* Fig. 12). Monoclonal antibodies are measured at low concentration in first-in-human applications but also at higher concentrations following the pharmacokinetic decay when higher doses are given. The most sensitive assay was obtained using Gyrolab Bioaffy 4000 with minimum required dilution (MRD) of 4 to give a broad analytical range with an approximate lower limit of quantification (LLOQ) of 6 ng/mL and upper limit of quantification (ULOQ) of 60,000 ng/mL in neat serum. This assay can be used for a broad range of analyte concentrations.

3.4 Biomarker Analysis: Interleukin 2 (IL-2)

3.4.1 Background

Biomarkers such as cytokines are widely measured to obtain information on a biological condition, examine a disease state, or as a response to a therapeutic drug (pharmacodynamics). Cytokines are used as indicators of systemic inflammatory response syndromes triggered by a variety of reasons. This includes cytokine storms

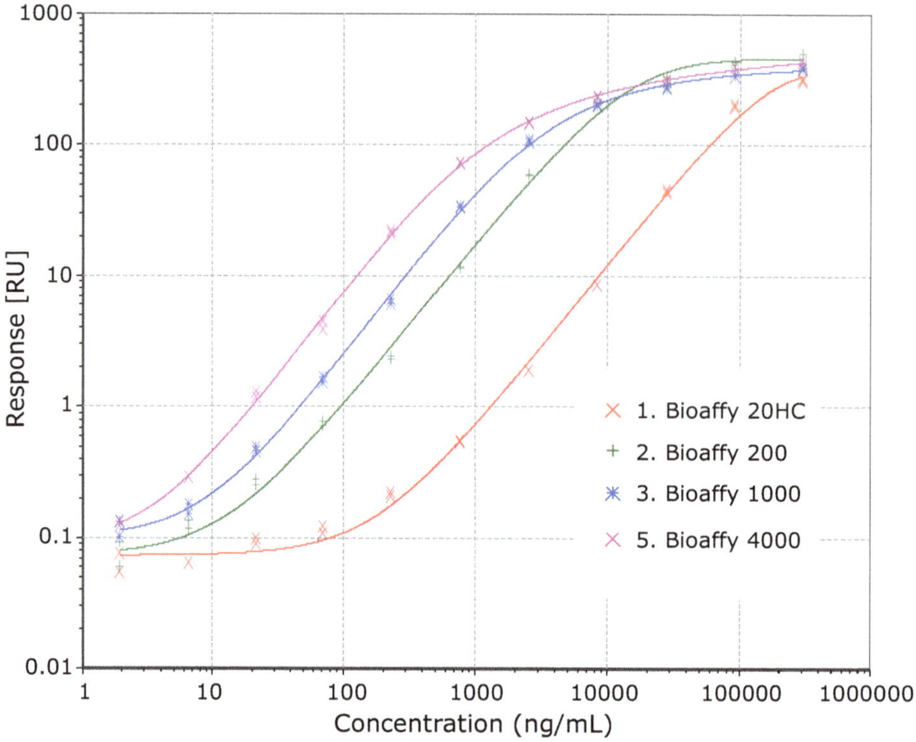

Fig. 12 A broad dynamic range was obtained using different Gyrolab Bioaffy CDs for the pembrolizumab bridging assay. The sensitive Gyrolab Bioaffy 4000 CD can be used to measure concentrations in the range of 10–4000 ng/mL. With Gyrolab Bioaffy 20HC, the upper limit of quantitation was extended to 100 μg/mL

associated with infections such as COVID-19 [4], adverse cytokine release syndrome resulting from treatment with therapeutic antibodies [5], and during chimeric antigen receptor T cell (CART) therapy used in treating cancer [6]. Cytokines and their receptors are also used as drug targets in the treatment of a variety of disorders such as cancer or autoimmune diseases. Biomarker applications require assays with high sensitivity.

Cytokines are also used as therapeutics, and analytical support in this development process benefits from assays with a broad dynamic concentration range that includes high-titer assays. These applications include pharmacokinetics assays and monitoring concentration throughout the development phases, including selection and optimization of cell lines for productivity and downstream purification steps.

Interleukin-2 (IL-2) has been selected as a model system to illustrate different analytical aspects of biomarker and cytokine therapeutics. IL-2 has had a profound impact on the development in cancer immunotherapy, and it was one of the first cytokine drugs approved [7]. With the increased activity in immunotherapy, there has been renewed interest in cytokines recently. This includes

Table 2
Estimated assay range, based on three runs in 50% human serum

Assay range	LOD (pg/mL)	LLOQ (pg/mL)	ULOQ (pg/mL)
On plate	~0.3	~1	~3000
In neat matrix	~0.6	~2	~6000

The limit of detection (LOD) represents the concentration corresponding to the assay blank +2 standard deviations

Table 3
Accuracy and precision data of QC samples in Rexxip H (n = number of runs)

QC	Expected concentration (pg/mL)	Average measured concentration (pg/mL)	Inter-run CV (%; $n = 3$)	Average intra-run CV (%; $n = 3$)	Average total error (%; $n = 3$)
1	1	1.1	17	13	24
2	2	2.1	11	11	18
3	5	5.0	5.9	2.4	6.7
4	10	10	10	1.8	8.7
5	100	104	1.5	3.5	7.4
6	3000	3001	2.4	3.3	5.0

combination of the PD-1 inhibitors, pembrolizumab and nivolumab, with high-dose PEGylated IL-2 to improve the antitumor effects [8, 9].

3.4.2 Assay Performance

A three-step sandwich assay was developed to quantify IL-2 over a concentration range of >6-logs using different Gyrolab Bioaffy CDs (*see* Fig. 12). The most sensitive assay was obtained with Gyrolab Bioaffy 4000 CD. A robust 4-log standard curve was generated with this CD, achieving an assay range of 1–3000 pg/mL in human serum diluted 1:2 (*see* Table 2). Accuracy and precision data for the IL-2 assay include quantification of QC samples with precision better than 20% CV (*see* Table 3). In the selectivity assessment, IL-2 spiked into human serum in a serial dilution resulted in recoveries in the range of 90–110%.

The other Gyrolab Bioaffy CDs have extended ULOQ including μg/mL levels using the Bioaffy 200 or Bioaffy 20HC. The CDs with high-capacity particles have narrower dynamic ranges due to increased assay background (*see* Fig. 13).

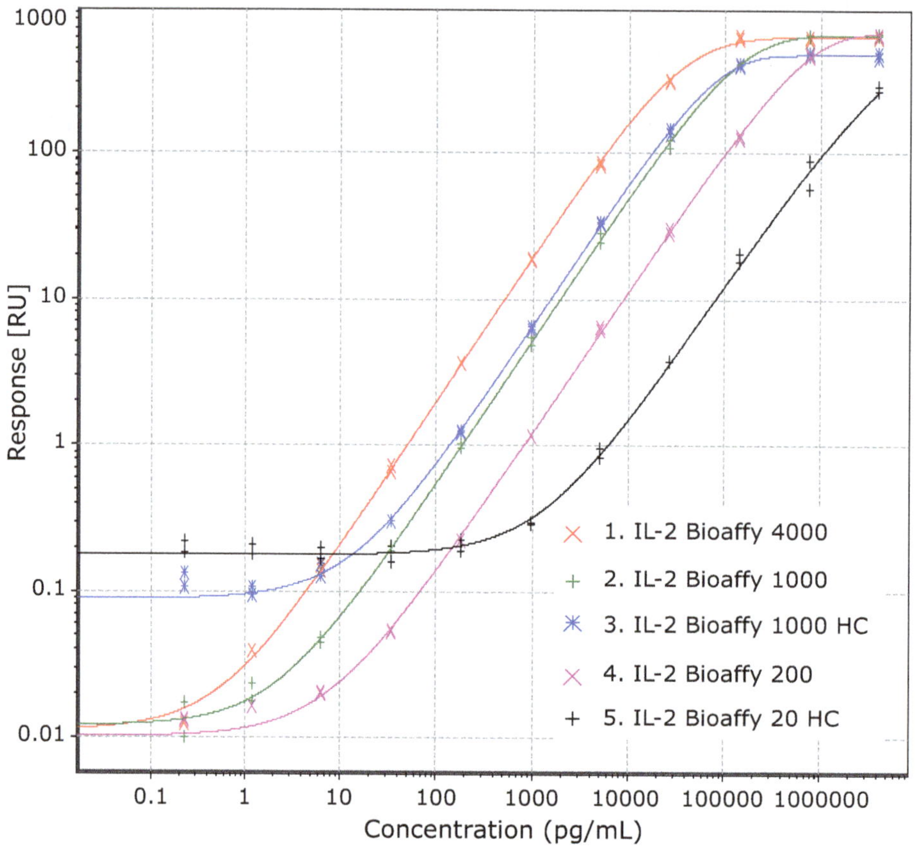

Fig. 13 IL-2 assay with dynamic range from 1 pg/mL to mg/mL using different Gyrolab Bioaffy CDs. Sensitive assays were obtained with Bioaffy 4000 while Bioaffy 200 or 20HC could be used for higher analyte concentrations. All assays have been performed with the same affinity reagents and concentrations using a three-step assay

4 Conclusions

Gyrolab immunoassay system automates many of the manual steps involved in an immunoassay, and assay protocols are incorporated into the Gyrolab methods to simplify the optimization process. This reduces manual handling and errors to improve assay reproducibility and robustness and also smoothens the transfer of assays between labs.

Assay performance is influenced by the affinity of the reagents, sample volume, and assay conditions, and performance can be readily adjusted depending on the analytical needs. Benefits when using the system include.

Saving time, sample, and reagents

- The miniaturized format reduces sample and reagent consumption.
- The flow-through assay format reduces assay time.

- Time and cost are reduced due to automation.
- Assay optimization can be quickly performed in a few experiments.

Achieving high sensitivity and broad dynamic range

- The fluorescence detection gives a broad dynamic range for each CD (generally 3–4 logs).
- The same assay conditions can be used for different Bioaffy CDs to meet different analytical needs and can extend the dynamic range to 6 logs.
- Low analyte concentrations (sub-pg/mL) can be measured with Gyrolab Bioaffy 4000 combined with high-affinity reagents.
- Gyrolab CDs including the HC particle can be used for reagents with low-affinity or polyclonal antibodies.

As a result, Gyrolab immunoassay platform can meet a wide range of analytical needs throughout the biotherapeutic workflow. Applications range from high-sensitivity biomarker analysis in discovery through PK analysis in the clinical phases to high-titer measurements in process development and manufacturing. Other applications that particularly benefit from the integrated workflow and dedicated software tools include impurity analysis (host cell proteins, Protein A, etc.), immunogenicity assessment, and affinity characterization.

References

1. Engvall E, Perlmann P (1971) Enzyme-linked immunosorbent assay (ELISA). Quantitative assay of immunoglobulin G. Immunochemistry 8:871–874
2. Andersson P, Jesson G, Kylberg G et al (2007) Parallel nl microfluidic analysis system. Anal Chem 79:4022–4030
3. Centanni M, Moes DJAR, Trocóniz IF et al (2019) Clinical pharmacokinetics and pharmacodynamics of immune checkpoint inhibitors. Clin Pharmacokinet 58:835–857
4. Hansel T, Kropshofer H, Singer T et al (2010) The safety and side effects of monoclonal antibodies. Nat Rev Drug Discov 9:325–338
5. Murthy H, Iqbal M, Chavez JC et al (2019) Cytokine release syndrome: current perspectives. Immunotargets Ther 8:43–52
6. Murthy H et al (2019) Cytokine Release Syndrome: Current Perspectives. ImmunoTargets and Therapy 8:43–52
7. Punnonen J, Rosen D, Zuniga L et al (2019) Cytokine therapeutics in cancer immunotherapy: design and development. Curr Pharmacol Rep 5:377–390
8. Buchbinder EI, Dutcher JP, Daniels GA et al (2019) Therapy with high-dose Interleukin-2 (HD IL-2) in metastatic melanoma and renal cell carcinoma following PD1 or PDL1 inhibition. J Immunother Cancer 7(1):49
9. Diab A, Tannir NM, Bentebibel SE et al (2020) Bempegaldesleukin (NKTR-214) plus Nivolumab in patients with advanced solid tumors: phase I dose-escalation study of safety, efficacy, and immune activation (PIVOT-02). Cancer Discov 10:1158–1173. Cytokine storms: understanding COVID-19. Nilam Mangalmurti and Christopher A. Hunter. Immunity 53, July 14, 2020

Chapter 10

Assessment of Immunologically Potent Carbohydrates

Denong Wang

Abstract

Carbohydrates have a great potential in generating structural and immunological diversities. Microbial pathogens often decorate their outmost surfaces with specific carbohydrate signatures. Carbohydrate antigens also differ significantly from protein antigens in physiochemical properties, especially in surface display of antigenic determinants in aqueous solutions. Technical optimization or modifications are often needed when we apply standard procedures for protein-based enzyme-linked immunosorbent assay (ELISA) to assess immunologically potent carbohydrates. We present here our laboratory protocols for performing carbohydrate ELISA and discuss several assay platforms that may be applied complementarily to explore the carbohydrate moieties that are critical for host immune recognition and induction of glycan-specific antibody responses.

Key words Antiglycan antibodies, Carbohydrate antigens, Carbohydrate ELISA, Glyco-conjugates, Glycolipids, Glycoproteins, Polysaccharides

1 Introduction

Carbohydrate antigens include carbohydrate-containing macromolecules of multistructural configurations, such as polysaccharides, glycolipids, glycoproteins, and synthetic glyco-conjugates [1–3]. Recognition of carbohydrates as antigens began with the study of microbial polysaccharides. In 1917, Dochez and Avery [4] found that when *Pneumococci* were grown in fluid media there was a substance in the culture fluid that precipitated specifically with antisera to the same *Pneumococcus*. Heidelberger and Avery [5] showed that this substance was a polysaccharide and not a protein, as previously thought. It was later found that almost every microbial pathogen expresses such "glyco-codes" that are recognized by the host immune systems and are effective in stimulating specific antibody responses [1, 6].

Carbohydrate moieties are unsurpassed in generating structural diversity and are prominent in surface display [1]. Unlike proteins, which are connected solely by a peptide bond, carbohydrates utilize

Robert S. Matson (ed.), *ELISA: Methods and Protocols*, Methods in Molecular Biology, vol. 2612,
https://doi.org/10.1007/978-1-0716-2903-1_10,

many possible glycosidic linkages so as to extensively diversify their structures. Two amino acid residues, such as two alanines, can produce only one possible dipeptide; however, two molecules of glucose have the potential to generate 11 different disaccharides. A trimer of any of the nine common sugar residues of the human body theoretically can give rise to 119,736 different structural isomers. This is strikingly in contrast to the maximal construction of 8000 tripeptides using 20 different amino acid residues. Theoretically, carbohydrates can have unlimited variation. This potential makes them well suited to provide biosignals for molecular and immunological recognition.

Polysaccharide antigens differ from protein antigens in many structural and immunological characteristics. Notably, a microbial polysaccharide may display both the terminal sugar moieties and internal chain structures on its solvent-accessible surface for immune recognition (Fig. 1). In aqueous solution, proteins tend to fold to bring their hydrophobic side chains together, forming an oily core with polar side chains exposed. Surface moieties of a protein antigen may serve as antigenic determinants interacting with B-cell Ig receptors; interior residues are generally not accessible to such interactions. By contrast, polysaccharides are built up by

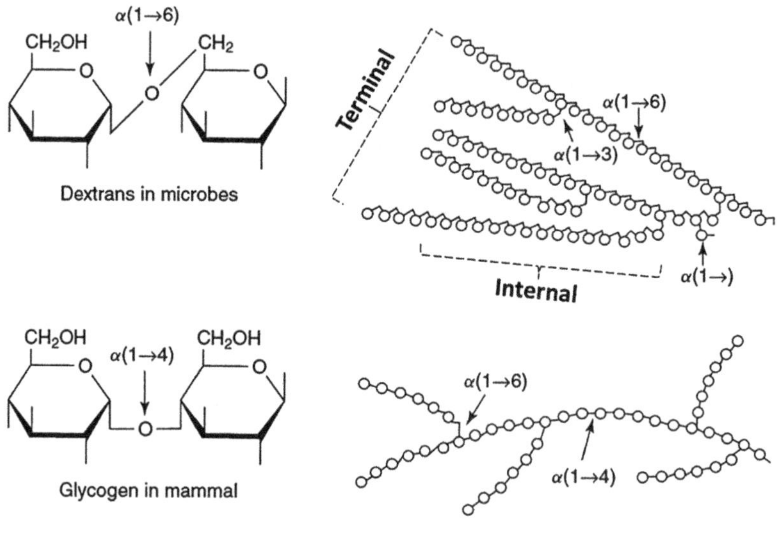

Fig. 1 Schematics of microbial dextran and mammalian glycogen: glycosidic linkages make the difference. The only source of structural diversity in homopolysaccharides is the glycosidic bonds linking the monosaccharides. This figure illustrates two commonly found polysaccharides—one in microbes and another in mammals. The glucosyl polysaccharide with $\alpha(1 \rightarrow 6)$ linkage is dextran, which is found in many bacteria, including *Lactobacillaceae*. The glucose polymers with $\alpha(1 \rightarrow 4)$ linkage predominating are found in mammals (e.g., glycogen) or plants (e.g., starch) as major types of the biological energy-storage molecules

monosaccharides whereby the enriched hydroxy groups readily interact with water molecules by hydrogen bonding. Their glycosidic linkages are more flexible than the peptide bonds in proteins and protein-like folding patterns are not seen in polysaccharides. Thus, not only are the terminals of the carbohydrate chains accessible for molecular recognition but residues in the internal chain are also exposed in solvent and are frequently immune reactive.

Carbohydrate ELISA has been proven to be highly valuable in diagnosis of microbial infection and assessment of vaccine-elicited antiglycan antibody responses [7–14]. We describe here our routine laboratory protocols for performing polysaccharide ELISA and illustrate several commonly used ELISA designs for identification and characterization of immunologically potent carbohydrates.

2 Materials

1. ELISA microtiter plates: The ELISA plates designed for coating protein antigens, such as Nunc-MaxiSorp-96-well plate and Immulon® Microtiter™ 96-Well Plates and Strips (Thermo Scientific), are applicable for polysaccharide ELISA.
2. Polysaccharide coating buffer: 0.02 M borate-buffered saline (BBS) pH 8.0 [1 L of 1× BBS is composed of 0.2 M borate acid (Sigma) 70 mL, 0.2 M sodium borate (Sigma) 30 mL, and sodium chloride 7.5 g] (*see* **Note 1**).
3. Blocking solution: 1% bovine serum albumin (BSA), 0.025% thimerosal, and 0.05% Tween 20, in 1× PBS (10 mM phosphate buffer pH 7.4, 150 mM NaCl).
4. Washing solution: 1× PBS with 0.05% Tween 20 (PBST).
5. Detecting agents: Alkaline phosphatase (AP)-, peroxidase (HRP)-, or biotin-conjugated secondary antibodies.
6. AP substrate: Nitrophenyl phosphate disodium salt hexahydrate (pNPP) tablets.
7. HRP substrates: 3,3′, 5,5′ tetramethylbenzidine (TMB) substrate reagent; o-phenylenediamine dihydrochloride (OPD) tablets.
8. Reaction-stopping reagents: 3 M NaOH for AP substrate; 3 M HCl or 2 M H_2SO_4 for HRP substrate.
9. Microtiter plate readers: Plate readers shall be equipped with 405 nm and 450/492 nm filters for AP and HRP substrates, respectively.

3 Methods

3.1 Antigen Coating and Blocking

1. Dilute a purified polysaccharide antigen in 1× BBS (pH 8.2) or a glycoprotein in 0.1 M bicarbonate buffer (pH 9.6) at the appropriate concentration (e.g., 5.0–10.0 μg/mL).
2. Pipette 0.1 mL of the above solution to each well of a 96-well microtiter plate.
3. Incubate at 37 °C for 2 h or incubate overnight at 4 °C.
4. Discard the coating solution and wash the plate with 0.2 mL PBST three times.
5. Add 0.2 mL 1% BSA PBST per well and incubate at room temperature for 1.0 h to block the antigen-noncoated reactive areas of each well.

3.2 Primary Antibody Reaction

1. Dilute the primary antibody preparation in 1% BSA PBST (*see* **Notes 2** and **3**).
2. Discard the blocking buffer and add 0.1 mL of the diluted primary antibody to each well and incubate the plate at 37 °C for 1.5 h.
3. Wash the plates with 0.2 mL1× PBST per well five times.

3.3 Secondary Antibody Reaction

1. Dilute the enzyme-conjugated secondary antibody in 1% BSA PBST. The optimal dilution should be determined using a titration assay.
2. Add 0.2 mL of this solution to each well.
3. Incubate at 37 °C for 1 h.
4. Wash the plates as in Subheading 3.2.

3.4 Substrate Preparation and Development

AP Substrate

1. During the last incubation and immediately before use, prepare the AP substrate by adding one pNPP tablet in 5 mL 1M diethanolamine solution pH 9.8 and bring the premade liquid substrate to room temperature.
2. Add 0.1 mL of substrate to each well. Yellow color should develop in positive wells after 30 min.
3. Read absorbance in a microplate reader at 405 nm at 30 and 60 min, respectively.
4. Stop the reaction with 50 μL per well of 3 M NaOH in 60 min before reading when multiple plates are used in the same experiment.

HRP Substrate

1. Immediately before use prepare the TMB HRP substrate by mixing equal amounts of solution A and solution B in the TMB substrate reagent set of BD Biosciences, Cat# 555214.
2. Add 0.1 mL of the freshly prepared substrate to each well. Green color should develop in positive wells in a few minutes.
3. Incubate the reaction at room temperature for 20 min.
4. Stop the reaction with 50 μL per well of the 2 M H_2SO_4.
5. Read absorbance in a microplate reader at 450 nm.

For other AP or HRP substrate, please follow the manufacturer's specific instruction.

3.5 Examples of Application

3.5.1 Indirect ELISA

This assay immobilizes a carbohydrate antigen in an ELISA plate for detecting antibodies or other glycan-binding agents in a solution. Since the antigen-captured antibodies are detected by a secondary agent, this assay is termed indirect ELISA. We may apply a well-characterized carbohydrate antigen to detect and quantify unknown antiglycan antibodies in a bodily fluid, such as a blood sample obtained from an infected or vaccinated subject. Such carbohydrate ELISA has been widely used in clinical diagnosis of microbial infection or for monitoring the efficacy of a carbohydrate vaccine, such as a microbial polysaccharide or a synthetic glycoconjugate that displays a protective glyco-determinant derived from a microorganism. We may also use antibodies of defined glyco-epitope-binding specificity to verify whether an immobilized carbohydrate antigen preserve critical antigenic determinants for a desired diagnostic application.

Figure 2 shows an ELISA analysis of dextran polysaccharides in comparison with a carbohydrate microarray analysis [3]. The goal of this study was to investigate whether immobilized carbohydrate macromolecules preserve their antigenic determinants or epitopes on corresponding substrate for antibody recognition.

As shown in Fig. 2a (left), a groove-type mAb, 4.3F1 bound to the dextran preparations with predominantly linear chain structures, N279 and LD7, but bound poorly to the heavily branched α(1,6)dextran, B1299S. By contrast, when the cavity-type mAb 16.4.12E (Fig. 2a, right) was applied, it bound to the immobilized dextran preparations having branches (N279 and B1299S) but not those with only internal linear chain structure (LD7). These patterns of antigen–antibody reactivities are characteristically identical to those recognized by an ELISA-binding assay (Fig. 2b) for either the groove type or the cavity type of anti-dextran mAbs. Thus, dextran molecules were immobilized on both nitrocellulose-coated glass slide (in saline) and ELISA plate (in BBS) and have their immunological properties well preserved. Both their nonreducing end structure, recognized by the cavity-type anti-α(1,6) dextrans,

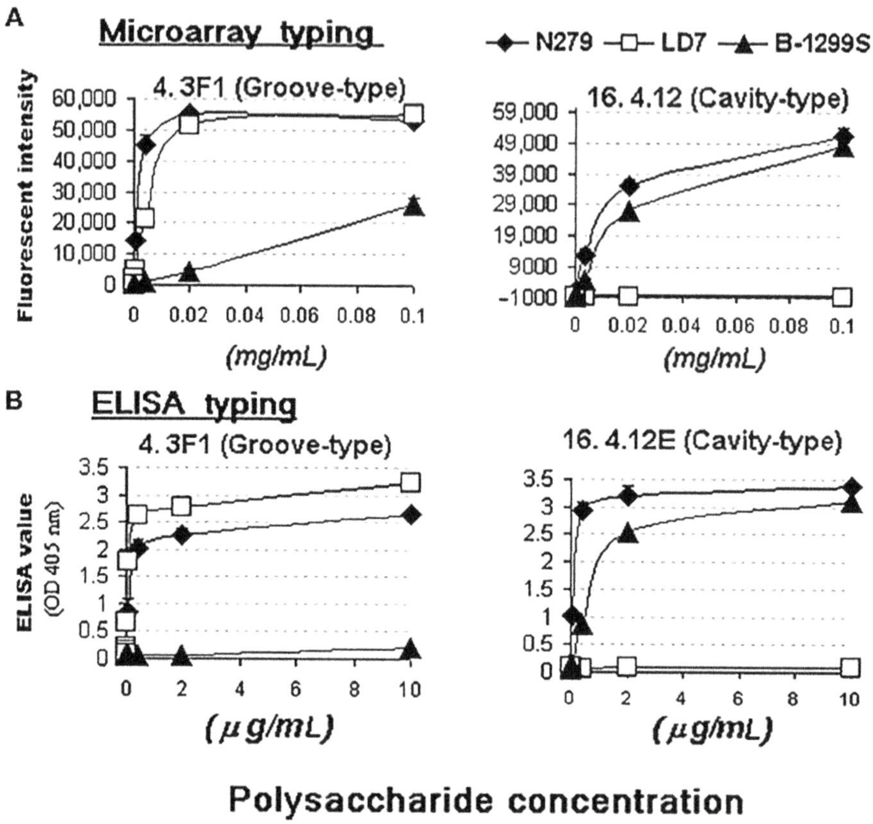

Fig. 2 Immunological characterization of surface-immobilized dextran molecules. The graphs were adapted from [3] with minor changes. Microbial polysaccharides dextrans of different linkage compositions and of different ratios of terminal to internal epitopes were used. These dextran preparations include N279, displaying both internal linear and terminal nonreducing end epitopes (Fig. 1); B1299S, heavily branched and expressing predominantly terminal epitopes; and LD7, a synthetic dextran composed of 100% α(1,6)-linked internal linear chain structure. The dextran preparations immobilized in microarray substrate or on ELISA plates were incubated with monoclonal antibodies (mAbs) of defined specificities, either a groove-type anti-α(1,6)dextran 4.3F1 (IgG3) or a cavity-type anti-α(1,6)dextran 16.4.12E (IgA). The former recognizes the internal linear chain of α(1,6) dextrans; the latter is specific for the terminal nonreducing end structure of the polysaccharide

and the internal linear chain epitopes, bound by the groove-type anti-α(1,6) dextrans, are displayed on the surface after immobilization and are accessible to antibodies in an aqueous solution.

3.5.2 ELISA Competition Assays

ELISA competition assay is often used to characterize the fine specificity of a glycan-binding agent, either an antiglycan antibody or a lectin. Experimentally, these assays examine whether a set of saccharides of different sequences and structural characteristics competitively inhibit a specific pair of carbohydrate–anticarbohydrate antibody interaction and thereby identify the saccharides that maximally interfere with or block binding of a carbohydrate antigen by an antibody preparation.

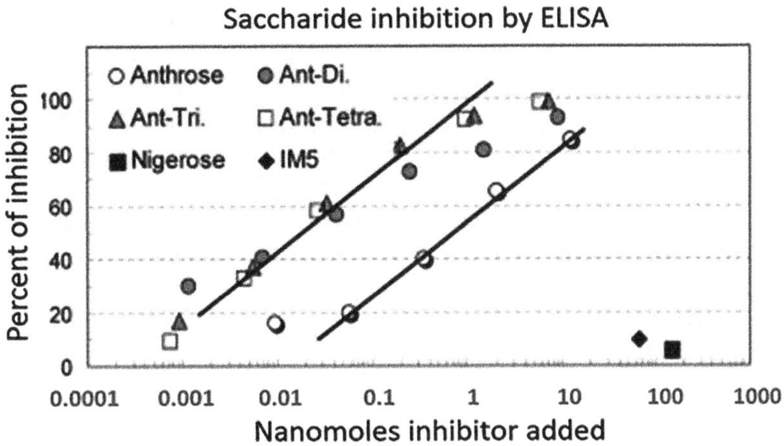

Fig. 3 ELISA-based quantitative saccharide inhibition assays. ELISA microtiter plates (NUNC, MaxiSorp) were coated with a BSA conjugate of a synthetic saccharide of a-anthrose-tetrasaccharide (5 μg/mL) were incubated with a preparation of rabbit antianthrax spore IgG (2 μg/mL) in the presence or absence of varying quantities of the sugar inhibitors. Percent inhibition was calculated as follows: % inhibition = ((standard A − blank A) − (A with inhibitor − blank A))/(standard A − blank A). The half-maximal inhibitory concentration (IC_{50}) values for the given saccharides were calculated based on mathematical models of the linear range of the corresponding saccharide inhibition curve. (Adapted from [8] with minor changes)

Figure 3 shows an effort to identify immunogenic sugar moieties expressed by *Bacillus anthracis* spores [8]. In this experiment, an anthrose–tetrasaccharide–BSA conjugate was coated on an ELISA plate to react with a preparation of rabbit antianthrax spore IgG antibodies. A panel of saccharides, including anthrose monosaccharide and anthrose-containing di-, tri-, and tetrasaccharides, were applied to competitively inhibit the antibody–glyco-conjugate interaction.

As illustrated, the quantitative inhibition curves generated by the four saccharides are nearly linear. The curve of the anthrose–trisaccharide (IC_{50}, 0.016 nmol) is essentially superimposed to the curve of the anthrose–tetrasaccharide (IC_{50}, 0.016 nmol). Both marginally differ from the disaccharide (IC_{50}, 0.019 nmol) but are significantly different from the monosaccharide (IC_{50}, 0.688 nmol). The relative inhibiting powers of the four anthrose saccharides are, thus, 1.00(tetra)/1.00(tri)/1.18(di)/43.3 (mono). In striking contrast, an α(1,3) glucosyl disaccharide, nigerose, and an α(1,6) glucosyl pentasaccharide, IM_5 (isomaltopentose), show no inhibition to the specific antibody reactivities with the anthrose–tetrasaccharide in the same assay.

This investigation demonstrates that rabbit IgG antibodies elicited by *Bacillus anthracis* spores specifically recognize a tetrasaccharide chain that decorates the outermost surfaces of the

B. anthracis exosporium. Given the significant inhibition potency of anthrose monosaccharide and disaccharide, a dominant antigenic determinant recognized by the rabbit antisera appears to be the terminal nonreducing epitopes of the *B. anthracis* exosporium oligosaccharide. The potential of this anthrose–tetrasaccharide as a potent immunological target for detection of *B. anthracis* spores and development of novel vaccines that target anthrax spores is yet to be further investigated.

3.5.3 ELISA "Sandwich" Assays (Capture ELISA)

A sandwich ELISA is usually supported by a pair of antigen-specific agents, one immobilized on the ELISA plate for capturing a substance in the solution phase and another as detecting agent for detecting and quantifying the captured substances. Such immunological "double authentication" assay design significantly improves detection specificity and sensitivity of sandwich ELISA.

An effective approach for developing a sandwich ELISA is to identify a pair of antibodies that recognize different epitopes of a given antigen. For example, a cavity-type anti-α(1,6)dextran mAb 16.4.12E (specific for the terminal nonreducing epitopes) may be used in combination with a groove-type mAb 45.21.1 (specific for the internal chain epitopes) to detect α(1,6)dextran polysaccharides. Figure 4 shows that a sandwich ELISA was applied to detect dextrans in the potato tubers that were genetically engineered to

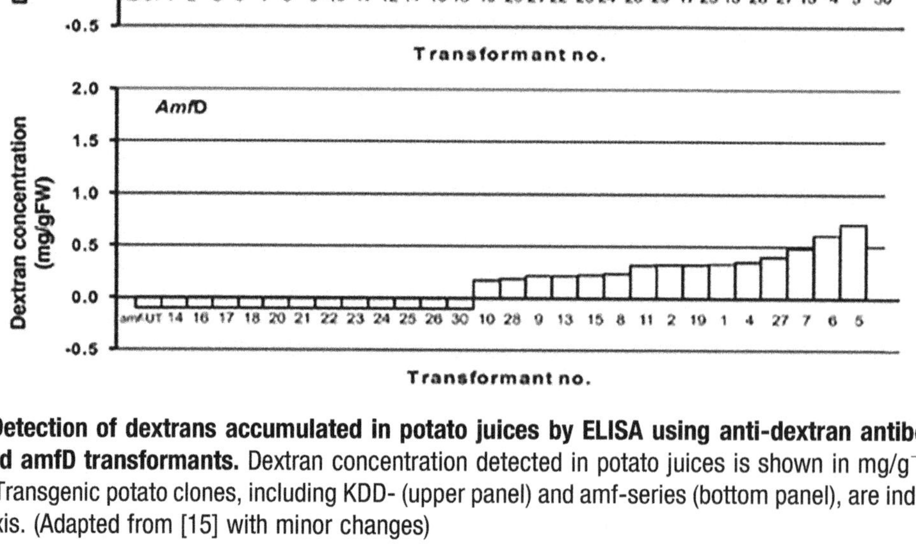

Fig. 4 Detection of dextrans accumulated in potato juices by ELISA using anti-dextran antibodies in KDD and amfD transformants. Dextran concentration detected in potato juices is shown in mg/g^{-1} FW in Y-axis. Transgenic potato clones, including KDD- (upper panel) and amf-series (bottom panel), are indicated in the X-axis. (Adapted from [15] with minor changes)

produce α(1,6) dextrans [15]. In the tuber juice of the KDD series, dextran was detected in 9 out of 30 tubers (29%) in a concentration ranging from 0.3 to 1.7 mg g^{-1} FW. In the amfD series, dextran was detected in 15 out of 27 tubers (56%), but the amount was lower (0.2 to 0.7 mg g^{-1} FW).

The goal of this potato bioengineering experiment was to produce human–noncleavable glucosyl polymer α(1,6)dextrans in potato so that the products are still edible but have lower calories compared to the native potato with the human-digestible starches as the main energy-storage molecules. A mature dextransucrase (DsrS) gene from bacteria *Leuconostoc mesenteroides* was expressed in two potato genotypes, Kardal (KDD) and the amylose-free (amf) mutant. Of note, dextrans and starches are *homopolysaccharides* of glucoses and differ only in their glycosidic bonds. This sandwich ELISA selectively detects one type of glucosyl polymers in a mixture of biological substances containing other types of glucosyl homopolymers.

It is noteworthy that bacterial polysaccharides are typically much more complex than those produced by plants or animals. They are often partially composed of unusual sugar residues that are rarely seen in higher eukaryotic species. Although only nine monosaccharides are commonly seen in mammals, over a hundred different monosaccharides are found in bacteria. Additionally, microbial polysaccharides contain various ester-linked substituents and pyruvate ketals, which are frequently of immunological significance. Given the structural diversity of carbohydrate antigens, it is essential to examine any new antigen preparation to determine its efficacy of immobilization and preservation of critical glycodeterminants in each type of ELISA plate to achieve optimized condition in a carbohydrate ELISA.

4 Notes

1. A bicarbonate buffer (0.1 M sodium carbonate/bicarbonate, pH 9.6) is often used to solubilize proteins and peptides and unprotonated them so that the antigens gain an overall negative charge for binding to a positively charged plate. Polysaccharides are generally hydrophilic and highly soluble in aqueous solutions, such as BBS and phosphate-buffered saline (PBS). We often use BBS for polysaccharide immobilization [3] and a bicarbonate buffer for glycoproteins and synthetic glycan–protein conjugates [16] in ELISA.
2. The optimal sample dilution should be determined using a titration assay. For detecting antiglycan antibodies in human blood samples, we often dilute serum or plasma at 1:500

dilution in blocking solution as a starting dilution for the initial screening and further titrate the samples for a tested carbohydrate antigen.

3. The assays using blood samples must be conducted in biosafety level (BSL)-2 (noninfected human blood samples) or BSL-3 (samples containing selected agent) facilities as specified in corresponding Biological Use Authorization (BUA) protocols.

Acknowledgments

We acknowledge the Kabat Collection of Carbohydrate Antigens at SRI International for a panel of carbohydrate antigens and antibodies that were applied in this study. This work was supported in part by the US Government grants R21DA046144-02 (NIH), 1R21CA254048-01 (NIH), and PR170128 (CDMRP) to DW. The content is solely the responsibility of the author and does not necessarily represent the official views of the funding agents.

References

1. Wang D, Kabat EA (1996) Carbohydrate antigens (polysaccharides). In: Regenmortal MHVV (ed) Structure of antigens. CRC Press, Boca Raton/New York/London/Tokyo, pp 247–276
2. Wang D (2004) Carbohydrate antigens. In: Meyers RA (ed) Encyclopedia of molecular cell biology and molecular medicine, vol II. Wiley, Weinheim, pp 277–301
3. Wang D, Liu S, Trummer BJ, Deng C, Wang A (2002) Carbohydrate microarrays for the recognition of cross-reactive molecular markers of microbes and host cells. Nat Biotechnol 20(3): 275–281. Epub 5 Mar 2002. https://doi.org/10.1038/nbt0302-275
4. Dochez AR, Avery OT (1917) The elaboration of specific soluble substance by pneumococcus during growth. J Exp Med 26:477–493
5. Heidelberger M, Avery OT (1923) The soluble specific substance of pneumococcus. J Exp Med 38:73–80
6. Mond JJ, Lees A, Snapper CM (1995) T cell-independent antigens type 2. Annu Rev Immunol 13:655–692
7. Ezzell JW Jr, Abshire TG, Little SF, Lidgerding BC, Brown C (1990) Identification of Bacillus anthracis by using monoclonal antibody to cell wall galactose-N-acetylglucosamine polysaccharide. J Clin Microbiol 28(2):223–231. Epub 1 Feb 1990. PubMed PMID: 2107201; PMCID: 269580
8. Wang D, Carroll GT, Turro NJ, Koberstein JT, Kovac P, Saksena R, Adamo R, Herzenberg LA, Herzenberg LA, Steinman L (2007) Photogenerated glycan arrays identify immunogenic sugar moieties of Bacillus anthracis exosporium. Proteomics 7(2):180–184
9. Schneerson R, Robbins JB, Barrera O, Sutton A, Habig WB, Hardegree MC, Chaimovich J (1980) Haemophilus influenzae type B polysaccharide-protein conjugates: model for a new generation of capsular polysaccharide vaccines. Prog Clin Biol Res 47:77–94. Epub 1 Jan 1980
10. Robbins JB, Schneerson R (1990) Polysaccharide-protein conjugates: a new generation of vaccines. J Infect Dis 161(5): 821–832. Epub 1 May 1990
11. Lucas AH, Rittenhouse-Olson K, Kronenberg M, Apicella MA, Wang D, Schreiber JR, Taylor CE (2008) Carbohydrate moieties as vaccine candidates: meeting summary. Vaccine 28(4):1121–1131
12. Granoff DM, Maslanka SE, Carlone GM, Plikaytis BD, Santos GF, Mokatrin A, Raff HV (1998) A modified enzyme-linked immunosorbent assay for measurement of antibody responses to meningococcal C polysaccharide

that correlate with bactericidal responses. Clin Diagn Lab Immunol 5(4):479–485. Epub 17 July 1998. PubMed PMID: 9665952; PMCID: PMC95603

13. Wernette CM, Frasch CE, Madore D, Carlone G, Goldblatt D, Plikaytis B, Benjamin W, Quataert SA, Hildreth S, Sikkema DJ, Kayhty H, Jonsdottir I, Nahm MH (2003) Enzyme-linked immunosorbent assay for quantitation of human antibodies to pneumococcal polysaccharides. Clin Diagn Lab Immunol 10(4):514–519. https://doi.org/10.1128/cdli.10.4.514-519.2003. Epub 11 July 2003. PubMed PMID: 12853378; PMCID: PMC164258

14. Inzana TJ, Champion A (2007) Use of an inhibition enzyme-linked immunosorbent assay for quantification of capsular polysaccharide or proteins in vaccines. Clin Vaccine Immunol 14(3):323–327. https://doi.org/10.1128/CVI.00302-06. Epub 3 Feb 2007. PubMed PMID: 17267591; PMCID: PMC1828856

15. Kok-Jacon GA, Vincken JP, Suurs LC, Wang D, Liu S, Visser RG (2005) Production of dextran in transgenic potato plants. Transgenic Res 14(4):385–395. Epub 6 Oct 2005. https://doi.org/10.1007/s11248-005-0439-0

16. Wang D, Dafik L, Nolley R, Huang W, Wolfinger RD, Wang LX, Peehl DM (2013) Anti-oligomannose antibodies as potential serum biomarkers of aggressive prostate cancer. Drug Dev Res 74(Biomarkers in drug development and companion diagnostics):65–80. https://doi.org/10.1002/ddr.21063

Chapter 11

Lateral Flow Microarray-Based ELISA for Cytokines

Robert S. Matson

Abstract

Cytokines are well known to be involved in numerous biological responses with diverse mechanisms of action, including the inflammatory process. The so-called "cytokine storm" has recently been associated with cases of severe COVID-19 infection.

Lateral flow microarray (LFM) devices have been constructed for multiplex detection of cytokines. The LFM-cytokine rapid test involves the immobilization of an array of capture anti-cytokine antibodies. Here, we describe the methods to create and use multiplex lateral flow-based immunoassays based upon the enzyme-linked immunosorbent assay (ELISA).

Key words Lateral flow, Microarray, ELISA, Cytokine, Multiplex, Rapid test

1 Introduction

In this chapter, we describe the steps in preparing lateral flow microarray (LFM) devices for the assessment of cytokines using a multiplex enzyme-linked immunosorbent assay (ELISA). The chapter is divided into sections that first describe the immobilization of capture antibodies by the "microarraying" process, followed by how to prepare the test strip and construct the LFM cassette. The chapter concludes with a protocol for the development and analysis of the rapid test for cytokines. The detection of human IL-6, EGF, and FGF-9 cytokines is used as an example.

1.1 Rapid Tests

The lateral flow immunoassay (LFIA) origin can be traced back to the application of paper chromatography by 1952 Nobel Laureates Martin and Synge (1944) and the development of the radioimmunoassay (RIA) first described by 1977 Nobel Laureates Yarlow and Berson in 1960. The immunochromatographic rapid test, now known as the lateral flow rapid immunoassay, has been successfully used as a diagnostic tool around the world for more than six decades [1]. Up until recently, the LFIA platform has been considered primarily as a qualitative test based upon a simple design

Robert S. Matson (ed.), *ELISA: Methods and Protocols*, Methods in Molecular Biology, vol. 2612,
https://doi.org/10.1007/978-1-0716-2903-1_11,

comprising a two-line visible indicator with an analyte detection zone or test (T) stripe and a control (C) stripe placed across a nitrocellulose membrane substrate. The home pregnancy test perhaps is the most recognizable example of the conventional LFIA rapid test [2].

However, advances in lateral flow test design, especially signal enhancements, have demonstrated improvements in sensitivity and quantification [3, 4]. Corstjens [5] reported the use of multiple test lines enabling the simultaneous detection of IL-10 and IFN-γ. In this early example of a multiplex lateral flow cytokine assay, there were two test lines and a single control line striped across the membrane. The first test line (T1) comprised anti-IL 10 monoclonal capture antibody (200 ng/stripe) set at 2.0 cm, followed by a goat antimouse control line (80 ng/strip) C1 at 2.5 cm. The anti-IFN-γ monoclonal capture antibody test line (T2) at 200 ng/strip was set at 5 mm upstream of the IL-10 capture antibody test line. The total length of the strip was 5 cm at a width of 4 mm. In this example, both cytokines achieved threshold in signal at 1 pg (equivalent to 100 pg/mL blood). The rapid assay was run at 10% serum load (10 μL serum + 90 μL LF buffer containing 100 ng each of anti-IL-10/anti-IFN-γ reporter conjugate) reaching threshold signal within 15 min. Swanson and D'Andrea [6] reported a similar multiplex line LFIA for IL-6 and C-reactive protein determination covering a range of 0–100 pg/mL IL-6 and 50–2500 ng/mL CRP, respectively, using the same strip.

1.2 Lateral Flow Microarray

Transformation of the lateral flow strip for multiplexing by placement of additional test lines is of limited utility. Zhao et al. [7] skirted the issue by creating a 10-channel lateral flow assay in which 10 individual test strips were arranged in a radial fashion within a disc. With a central shared port, sample could be applied, and 10 analytes simultaneously processed. Recently, He et al. [8] reported on the creation of multipath lateral flow strips using the laser direct-write (LDW) technique. Essentially, LDW photopolymerization creates multiple microchannels in the membrane such that a single strip can be partitioned with different test and control stripes for multiplexing. The flow paths thus allow simultaneous detection of multiple analytes.

A more conventional approach was first described by Carter and Cary [9] in which a single lateral flow membrane strip was printed with a microarray of oligonucleotide capture probes. They referred to this as the LFM system. Gantelius et al. [10] created an LFM for analysis of 384 protein fragments. A subset of 26 purified rabbit IgG isolates were analyzed in triplicate on the LFM protein system, and the results were found to be in good agreement with those immobilized on a glass microarray slide. Taranova et al. [11] applied LFM for the determination of samples for drugs of abuse. In this case, monoclonal capture antibodies were printed down in an array format to the nitrocellulose membrane, while anti-hapten (drug) antibodies were conjugated to colloidal gold particles (30 nm) separately and then applied to the conjugate pad as a mixture. The study compared the multiplex format to that of single analyte (stripped) LF strips. Both multiplex and single-plex tests achieved similar levels of accuracy and runtimes. The LFM version exhibited a higher sensitivity, e.g., amphetamine detection by LFM at 1.5 ng/mL and the singlet test at 5 ng/mL.

2 Materials

2.1 Immobilization of Anti-Cytokine Capture Antibody

1. Solid phase: Nitrocellulose membrane, polyester backed (Sartorius, CN95; *see* **Note 1**).
2. Lateral flow back slit card: 60 mm × 300 mm (Lohmann; DCN).
3. Capture antibodies: Purified, carrier-free, concentration ~ 0.5–1.0 mg/mL (*see* **Note 2**).
4. Print buffer: Phosphate-buffered saline (PBS), 0.01 M, pH 7.4 (Sigma-Aldrich P-3813).
5. Red dye marker: 3% red food dye in PBS, sterile filtered.
6. Biotinylated BSA marker: Sigma-Aldrich, A8549.
7. Filtration device: 0.2 μm sterile filtration of buffers.
8. Source plate: Nonbinding, 384-well (*see* **Note 3**).
9. Microarray dispenser: Capable of droplet dispense from ~1 to 10 nL (*see* **Note 4**).
10. Microcentrifuge: 0–2000 rpm.
11. Pipettor: Single-channel; multichannel.
12. Pipette tips: 1–10 μL; 100–1000 μL capacity.

2.2 Construction of LFM-Cytokine Cassettes (See Note 5)

1. Sample pad: GFDX, 10 mm × 300 mm (Millipore).
2. Conjugate pad: SureWick GFCP, 10 mm × 300 mm (Millipore).
3. Wick: SureWick CFSP, 17 mm × 300 mm (Millipore).
4. Cassette: The DCNovations Lateral Flow Cassette (DCN).

2.3 LFM-Cytokine Rapid Test

1. Biotinylated anti-cytokine antibodies: Concentration ~0.5–1.0-mg/mL (*see* **Note 2**).
2. Cytokine standards: Cytokine content certified; concentration, ng/mL (*see* **Note 2**).
3. Plasma samples: For analysis of cytokines (*see* **Note 6**).

4. LF (2×) diluent: 2% (w/v) PVA, 20% sucrose (w/v), 2% (v/v) Tween-20 with 2% (w/v) bovine serum albumin in PBS buffer, pH 7.5–8.5. Sterile filter and store at 4 °C (*see* **Note 7**).
5. LF chase buffer (*see* **Note 7**).
6. Streptavidin-HRP: Ready-to-use solutions preferred. Dilute as directed.
7. TMB, insoluble formulation: Ready-to-use precipitating enzyme (HRP) substrate (*see* **Note 8**).

2.4 Signal Development and Analysis (See Note 9)

1. Digital Microscope.
2. ImageJ software.
3. EXCEL spreadsheet.

3 Methods

3.1 Immobilization of Anti-cytokine Capture Antibody to Nitrocellulose Substrate

This protocol is for the preparation of capture antibody microarrays by direct adsorption to backed nitrocellulose strips (NC tape). In this case, anti-cytokine capture antibodies are prepared in a print buffer and dispensed to the surface using a microarray dispenser capable of delivering picoliter droplets. Nitrocellulose media is well known to sequester antibody protein while preserving their antigen-binding activity. To create a microarray on nitrocellulose, it is necessary to accurately dispense very small droplet volumes of each antibody to control the spreading of the spotted antibody to adjacent spots. Moreover, noncontact dispensing is required since contact printing with split-pin (quills) is easily fouled and will damage the membrane.

There are several steps required in constructing the LFM device. A back slit card is used for placement of the NC tape and other elements of the lateral flow strip. The NC tape is mounted to the back slit card for printing of the array (the Print Card). First, red dye is printed perpendicular down the mounted NC tape in parallel lines, called "cutlines." These are used to section the print card into the lateral flow strips. In the second step, the capture antibodies are arrayed between the cutlines. Finally, a control line is prepared by printing biotinylated BSA above the array. The strip sectioned along the cutlines is then mounted into the cassette after placement of the sample pad, conjugate pad, and wick on the print card (*see* Fig. 1).

3.1.1 Construction of the Nitrocellulose Print Card

1. Cut a section of the NC tape and secure it to a back split card to create the print card.
2. Cut the constructed print card to the required dimensions for printing.
3. Place the print card on the microarrayer deck and apply vacuum such that the print card remains flat and does not move out of position.

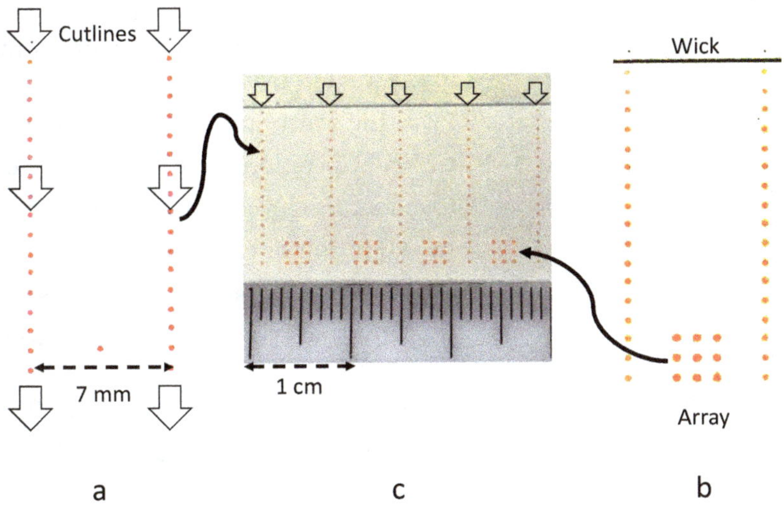

Fig. 1 Construction of the LFM strip. A piezo-dispenser (sciFLEXARRAYER, Model S5, Scienion AG) was used to print all features on the nitrocellulose (NC) strip. First, the NC strip was attached to a backing card and placed on the printer deck. Cutlines (**a**) were created using sterile filtered red food color dye prepared at 3% (v/v) in PBS, pH 7.4. Spots (500 pL × 20 drops or ~ 10 nL total droplet volume) were printed on a 1000-μm (1 mm) center-to-center pitch down the length of the nitrocellulose strip (25 mm) to create dotted lines spaced 7 mm apart across the membrane. An array of anti-cytokine antibodies (~36 nL total droplet volume) at a concentration of 0.25–0.5 mg/mL (9–18 ng/spot) was created near the lower end of the NC strip in a 3 × 3 pattern (**b**). This process was repeated across the backing card (**c**). Following completion of the printing, the backing card was removed from the deck and placed in a desiccator for drying. Once sufficiently dry, the sample pad and wick were placed on the backing card and the assembled LFM card sectioned into 5–6 mm strips. The completed strips were subsequently mounted into lateral flow cassettes for use

3.1.2 Printing

It is highly recommended that the user wear gloves during these procedures to avoid contamination issues and accumulation of debris on surfaces.

1. Prepare the antibodies for printing at 1 mg/mL in PBS, pH 7 (*see* **Note 10**).
2. Place 70 μL of each antibody solution in an assigned well of a 384-well source plate.
3. Place 70 μL of red dye marker in one well and 70 μL of biotinylated BSA marker in another well.
4. Put a lid on the plate.
5. Spin the plate at 2000 rpm for 5 min to remove air.
6. Set up the microarray dispenser for printing. Droplet volume should be adjusted to about 400–500 pL. Set the dispense to 20 droplets per spot or approximately 8–10 nL total spot

volume or about 8–10 ng protein. Critical conditions are relative humidity (RH), temperature, and dispense element cleaning. Set RH ~60%, temperature ~25 °C.

7. Place the source plate on the deck of the microarray dispenser.
8. Carefully position the nitrocellulose card on the print deck.
9. Conduct the programmed print run cycles (*see* **Note 11**):
   (a) Printing of cutlines.
   (b) Printing of capture antibodies.
   (c) Printing of biotinylated BSA control marker.

3.1.3 Storage

1. Remove the completed arrayed print card from the deck.
2. Place card in desiccator at room temperature for short-term storage (*see* **Note 12**).

3.2 Construction of the Lateral Flow Cassette

1. The print card is placed on the alignment fixture (*see* **Note 13**).
2. Remove the liner from the print card for placement of the conjugate pad. First, align the conjugate strip over the print card, then place it on the card such that a 2 mm overlap is made with the NC tape bottom edge. Press the conjugate strip firmly in place along the card.
3. Remove the liner from the print card for placement of the sample pad. Allow a 2 mm overlap with the bottom edge of the attached conjugate pad. Proceed to secure to the print card as provided in **step 2**.
4. Remove the liner from the print card for placement of the wick. Allow a 2 mm overlap with the top edge of the NC tape. Proceed to secure to the print card as previously described.
5. The completed print card is now ready for sectioning into individual strips by cutting along the printed cutlines.
6. The sectioned lateral flow strip is then assembled in the plastic cassette (*see* Fig. 1).
7. Return the completed LFM cassettes to the desiccator for storage.

3.3 Lateral Flow Microarray Cytokine Rapid Test

The LFM rapid test is run in a similar fashion to that of a standard lateral flow line assay. That is, the sample is added to the cassette and allowed to flow across the surface of the strip toward the wick. Analyte within the sample is likewise captured and detected during the process. However, as previously described in Subheading 1, there are some important differences with the multiplex rapid test. First, capture lines are replaced with spots formed from droplets dispensed to the surface using a high-precision noncontact dispenser. Figure 2 represents a schematic of the LFM rapid strip. The "test line" is replaced with the "test array" comprising

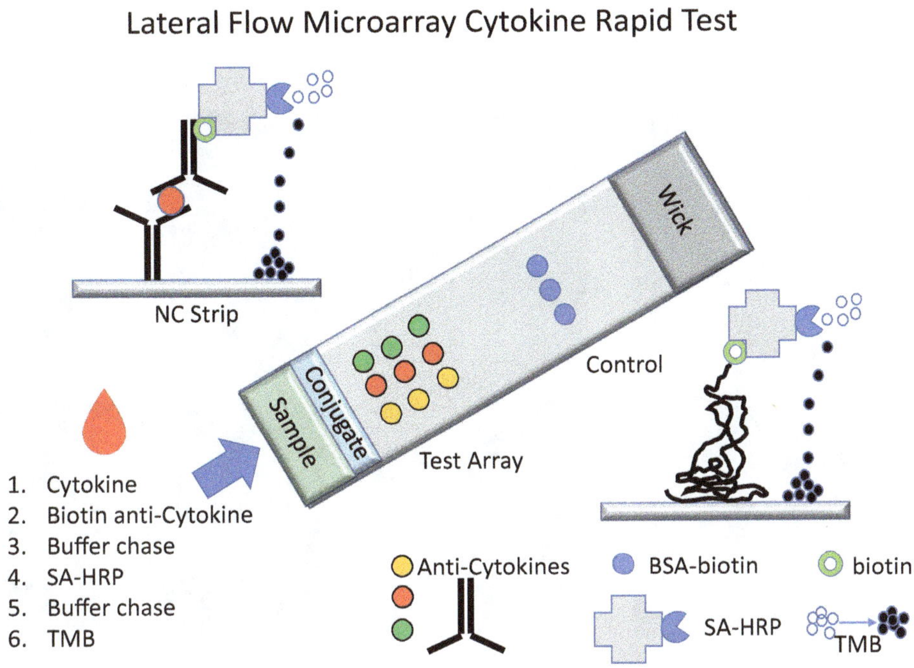

Fig. 2 Features of the LFM-cytokine rapid test. The LFM strip comprises a sample pad, conjugate pad (optional), NC membrane, and wick. The test array is made up of capture antibody spots corresponding to specific analytes of interest. The control spots are biotinylated BSA and serve to capture excess streptavidin-HRP to verify completion of fluid flow and signal development. The order of addition is as follows: sample (*1*) cytokine is mixed with (*2*) biotin anti-cytokine pool, (*3*) buffer chase, (*4*) addition of SA-HRP, (*5*) buffer chase from sample port of cassette, and finally, (*6*) insoluble TMB substrate is applied directly to the NC membrane for color development

individual spots of different capture antibodies. In this case, triplicate spots of three different anti-cytokine antibodies are immobilized closest to the conjugate pad in a 3 × 3 array pattern. Similarly, three spots are deposited upstream near the wick serve as a "control line." Here, biotinylated bovine serum albumin is immobilized to capture applied streptavidin reporter (HRP). To be clear, standard lateral flow rapid tests generally are not run as an ELISA. Instead, colloidal gold beads with immobilized secondary antibody are utilized for detection. To use this format for the multiplex, several sets of antibody-bound particles representing the different analytes would need to be pooled in proper proportion. This approach may work fine for a few analytes but is more taxing as the multiplex level increases.

The method described below is adapted from a conventional multiplex microwell ELISA. It is a stepwise process. First, sample is diluted in the presence of a pool of biotinylated secondary antibodies and applied to the sample port. The conjugate pad is removed because it is no longer necessary since no conjugates (gold particles) are employed. The fluid is allowed to reach the

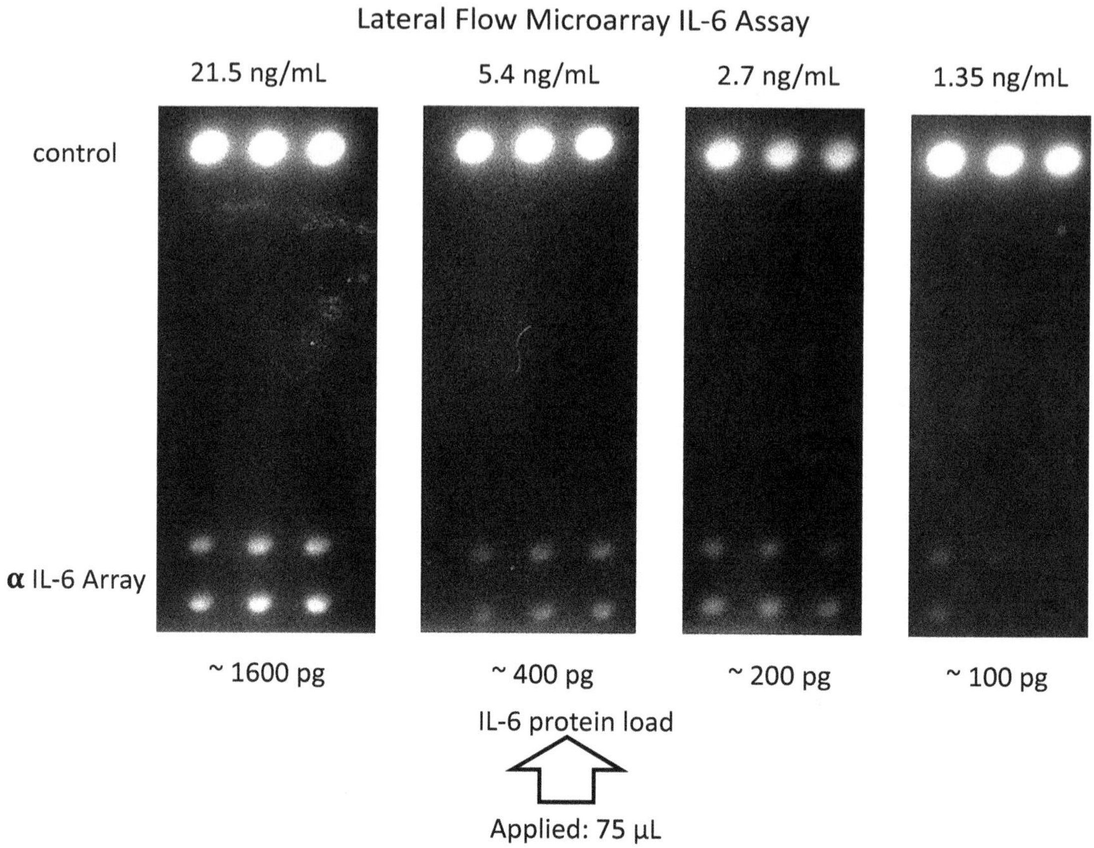

Fig. 3 Images of the LFM rapid test for human IL-6. Conditions: Anti-IL-6 capture antibody printed onto NC lateral flow membrane in 2 × 3 pattern (IL-6 Array); biotinylated BSA printed as reagent control (three spots). IL-6 standard protein (antigen) mixed with biotinylated (2°) antibody applied to sample pad. The IL-6 protein antigen was serially diluted 1:2 (v/v) in 1% BSA–PBS, pH 7.4; and 75 μL of each diluted sample applied to separate strips. Following clearance from the strip, biotin anti-IL 6 was applied, followed in order of addition: buffer chase, SA-HRP; buffer chase; TMB reagent (40 μL of each solution). Input concentrations (ng/mL) and calculated applied mass load (pg) per strip

wick and a buffer chase is used to clear membrane of sample. Streptavidin-HRP is then applied and allowed to migrate, followed by a second buffer chase. Finally, an insoluble TMB substrate is added to develop the signal. In our examples, we first demonstrate the analytical sensitivity of an LFM IL-6 ELISA (*see* Figs. 3 and 4), followed by the detection of additional cytokines (*see* Figs. 5, 6, and 7).

3.3.1 Performing the LFM-Cytokine Rapid Test

1. Place the LFM cassette on a flat surface.
2. Prepare a mixture (B) of biotinylated anti-cytokine antibodies in LF diluent.

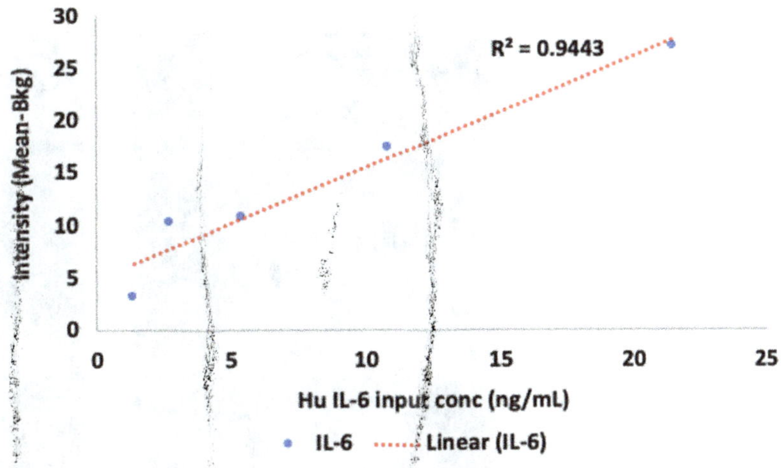

Fig. 4 Dose–response curve for human IL-6 determination. The mean pixel intensity of each developed spot was determined using ImageJ and corrected for mean background for the strip. The corrected spot intensity ($n = 6$ spots) vs. human IL-6 input concentration plotted as a linear regression ($R^2 = 0.9445$)

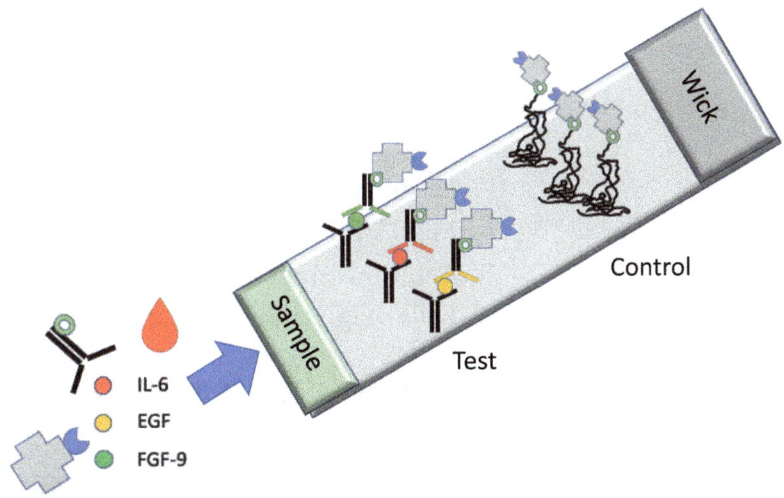

Fig. 5 Layout of the LFM-cytokine panel for IL-6, EGF, and FGF-9. Duplicate spots of anti-cytokine antibodies for human IL-6, EGF, and FGF-9 were printed in a 2 × 3 array pattern. Biotinylated anti-cytokine secondary antibodies (matched pairs) were pooled and mixed with the sample. The mixture was applied to the sample pad followed by a buffer chase. Once the fluid had migrated to the wick, streptavidin-HRP was applied. The resulting sandwich ELISA was then developed using an insoluble TMB reagent. Positive control for signal was determined using biotinylated BSA

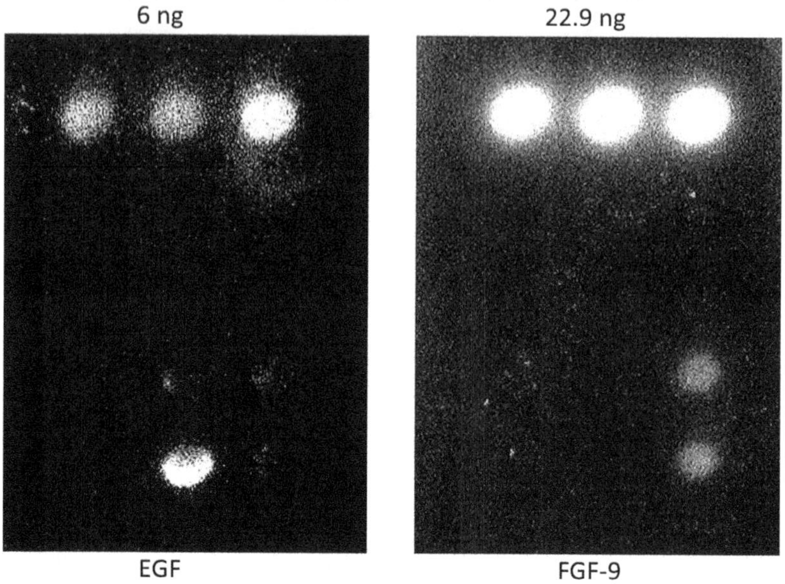

Fig. 6 Images of the LFM rapid test for human EGF and FGF-9. LFM-cytokine panel strips as described in Fig. 5 were used to detect the presence of individual cytokines under high loading conditions. For EGF, 6 ng (6000 pg) load of the cytokine applied followed by 40 ng biotin anti-EGF. For FGF-9, a load of 22.5 ng cytokine applied and detected using 72 ng biotin anti-FGF 9. In these experiments, IL-6 antigen or biotinylated anti-IL 6 antibody was not added

3. Prepare a mixture (S) of cytokine standards or plasma sample in LF diluent.
4. Mix 25 μL (S) + 50 μL (B).
5. Apply mix from **step 4** to the sample pad.
6. Following clearance from the strip apply 40 μL of LF chase buffer.
7. Apply 40 μL SA-HRP.
8. Apply 40 μL TMB.
9. Allow 15 min for signal to develop.

3.3.2 Signal Development and Analysis

1. Place cassette under digital microscope for viewing (*see* **Note 14**).
2. Adjust focus to allow "test array" and "control spots" to be observed within the field.
3. Capture image in monochrome and store as .jpeg or .tiff file.
4. Evaluate image by importing into ImageJ analysis program.
5. Invert the image color (dark to white spots) and change image type from RGB to 8-bit grayscale.

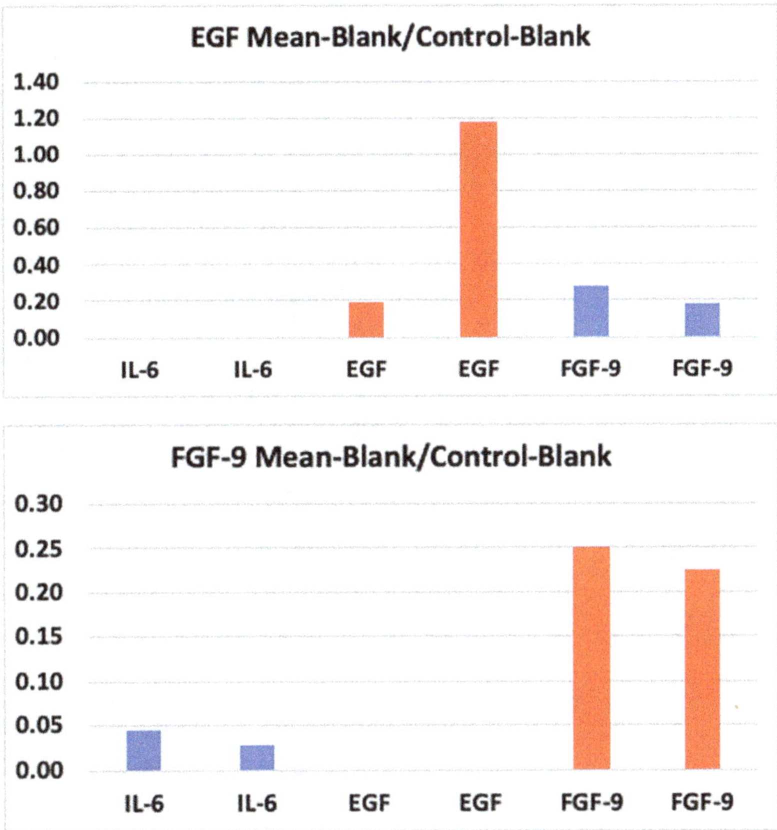

Fig. 7 Analysis of cross-reactivity for cytokines EGF and FGF-9. Analysis of the levels of cross-reactivity or nonspecific binding on the LFM-cytokine panel under high loading conditions (*see* Fig. 6). ImageJ was used to determine the pixel intensities of all spots corresponding to IL-6, EFG, and FGF-9. The biotinylated BSA control mean intensity was used to normalize analyte spot mean intensities. EGF showed low cross-reactivity with FGF-9, while FGF-9 exhibited a low cross-reactivity with IL-6 (*see* **Note 15**)

6. Using the spot analysis tool (circular ROI), measure the pixel intensity of the array spots, control spots, and selected ROI for background.
7. Transfer the Results.csv file into EXCEL for further analysis.

4 Notes

1. Backed nitrocellulose membrane from other manufacturers is available and may be suitable for use. The criteria for selection of the material include capillary flow rate, protein-binding capacity, thickness, and nonspecific binding. For example, Sartorius CN95 is considered a fast-flow material rated at 65–115 s/4 cm travel, while GE-Whatman FF80HP runs at

60–100 s/4 cm. There are slower capillary flow membranes that may improve resolution and binding. Also, different membranes may be required depending upon the sample matrix: blood, plasma, saliva, and urine. Evaluation of different materials for optimal performance is recommended.

2. Preparation of capture antibodies: The protocol calls for the development of a sandwich ELISA. If using monoclonals for this purpose, it is highly recommended that matched pairs of capture–reporter antibodies be used that recognize different epitopes of the antigen. It is advisable to screen at least two different sets of antibodies for sensitivity and specificity. In building a multiplex assay, it is also necessary to check for cross-reactivity, especially for assays employing polyclonal antibodies. Select well-characterized, purified antibodies. Certain antibody preparations may include stabilizing agents (carriers) such as albumin. Avoid these if possible since the carrier may compete with the antibody for immobilization to the nitrocellulose membrane. The antibody stocks should be maintained at a higher concentration to prevent denaturation. Keep these under refrigeration and avoid vigorous mixing that would cause denaturation. For printing, dilute the stocks into a suitable print buffer at various concentrations to determine optimal loading to the nitrocellulose that achieves significant antigen binding. Biotinylated anti-cytokine secondary antibody stock is to be maintained at higher concentration and stored refrigerated but not frozen. Dilute working stocks to about 1 μg/mL in LF diluent or 1:1000 (v/v) for stocks stored at 1 mg/mL. In pooling the different biotin anti-cytokines, try to maintain the final concentration of each into the range of 0.5–1 μg/mL. Cytokine standards that are supplied lyophilized should be diluted with a stabilizing buffer such as 1% BSA–PBS, pH 7.4, or that recommended by the vendor. Cytokines vary in their stability. Maintain stock solutions under refrigeration in the ng/mL range for 1–2 weeks. It is recommended that they be aliquoted and placed at −20 to −80 °C for long-term storage. Thaw once and discard after use. For multiplexing, the standards can be pooled as a standard mixture in stabilizing buffer and diluted to cover the range of about 10–1000 pg/mL. The standard mix may be used for about 1 week under refrigeration or aliquoted and stored for later use as previously described.

3. The source plate is used to store the diluted capture antibodies for printing. Polystyrene or polypropylene plates may be used for this purpose. However, it is important to verify and use either nonbinding or low-binding plates. Over time, significant loss of protein can occur from adsorption to plate wells that

have not been treated. Different proteins may have different rates of adsorption. These issues can lead to problems with print reproducibility.

4. Nitrocellulose membrane strongly adsorbs proteins such as albumins and immunoglobulins (e.g., capture antibody). However, the aqueous solutions used for printing contain buffers and wetting agents that tend to rapidly disperse upon contact with the surface. For creation of microarrays, it is important to control the spread of droplets to produce smaller protein spots with suitable morphology. This is best accomplished using a noncontact dispenser system capable of rapidly delivering picoliter (pL) to low nanoliter (nL) volumes to the surface.
5. Sample and conjugate release pads are generally a coated glass fiber while wicking pads are thicker absorbent cotton fiber materials. Here, wicking rate (s/4 cm), thickness (μm), and water adsorption (mg/cm^2) are important criteria for successful construction of the LFM device. If working with blood, a plasma separation membrane can be substituted for sample pad. The conjugate pad is not required for the LFM-cytokine ELISA cassette.
6. Proceed with caution while evaluating the level of cytokines in plasma! A recent study [12] concluded that "processing delays up to 15 h under refrigerated conditions are acceptable for the measurement of most circulating cytokines." However, there were exceptions, including significant differences between serum and plasma cytokine levels. This has been noted by numerous other studies as well. For doping of cytokine standards into normal plasma, we recommend EDTA-plasma as the matrix. Dilute the cytokine plasma at least 1:4 (v/v) in LF diluent before applying to the LFM cassette.
7. LF diluent (2×) was originally developed for stabilization and release of gold conjugates from the conjugate pad. It has been used successfully used at (1×) strength as a sample and reagent delivery medium for the LFM assay. When used as a chase buffer, the diluent may be used at full (2×) strength.
8. A precipitating TMB substrate is used to develop signal during the LFM assay. There are several formulations available: 1-Step Ultra TMB Blotting grade (Pierce #37574); TMB-M or TMB-MX (Moss, Inc.); and Bio-FX Enhanced (Surmodics).
9. The LFM assay produces an array of spots that while visible to the naked eye are best analyzed under magnification. An imaging reader such as a colorimetric scanner, microarray scanner, or digital microscope is recommended for analysis. A Dino-Lite Edge Digital Microscope, AM4115T-GFBW, was used for image capture of the LFM cassette array illuminated using a

blue-light LED in the described experiments. Image files were imported into ImageJ, which is a public domain software (https://imagej.nih.gov/ij/download.html), for the determination of pixel intensities and other properties of the spotted array.

10. For most applications, the printing of antibodies onto nitrocellulose can be accomplished using PBS buffer, pH 7.2–7.5. The print buffer should be sterile filtered. In order to prevent spot spreading issues, do not add additives to the buffer. Start with a protein concentration of about 1 mg/mL and serial dilute until the optimal antibody concentration for binding has been achieved.

11. A description regarding the construction of the LFM strip is provided in Fig. 1 in which a low-volume precision dispenser system was used to create the cutlines and microarray features. However, there are several manufacturers of such equipment, and each has its own unique programming formats. As a result, it is recommended that the reader contact the appropriate vendor for assistance in creating the LFM. The following offer high-precision dispensing: Scienion AG, Berlin, Germany (www.scienion.com); BioDot, Irvine, CA (www.biodot.com); ArrayJet, UK (www.arrayjet.co.uk); GeSiM mbH, Germany (www.gesim.de); and M2-Automation, Berlin, Germany (www.m2-automation.com).

12. After printing the LFM, it is important to completely dry down (cure) the immobilized proteins. This is best accomplished by placing the cards in a sealed dry box with desiccant for at least 1 day prior to use.

13. It is important that the lateral flow components (pads, NC strip, wick) are properly assembled onto the back slit card. This requires alignment and overlap of the components. Manual placement of materials to create the LFM could be problematic, and some sort of fixturing may be required. A vacuum lamination system, commonly called a "clam shell," can be employed to precisely overlay materials and seal. For example, BioDot (Irvine, CA) offers such a unit (LM5000) or Kinematic Association's Matrix laminator are suitable. Moreover, the sectioning of the card into individual LFM strips is greatly improved using a membrane "guillotine" cutter that provides high-quality precision cuts. These are commercially available from vendors such as BioDot or Kinematic, as well as others.

14. Imaging of the developed LFM cassette under a light source must be done without casting a shadow across the NC strip. Depending upon the light angle and intensity, the sides of the cassette can cast shadows down onto the strip. This is

particularly noticeable in using a high-intensity white light source. One solution is to remove the strip from the cassette for imaging.

15. During the development of a multiplex ELISA, it is critical to determine the nature and extent of any cross-reactivity or nonspecific binding since these will have a significant effect on both sensitivity and specificity [13].

References

1. Andryukov BG (2020) Six decades of lateral flow immunoassay: from determining metabolic markers to diagnosing COVID-19. AIMS Microbiol 6(3):280–304
2. Braunstein GD (2014) The long gestation of the modern home pregnancy test. Clin Chem 60(1):18–21
3. Liu Y, Zhan L, Qin Z et al (2021) Ultrasensitive and highly specific lateral flow assays for point-of-care diagnosis. ACS Nano 15:3593–3611
4. Zherdev AV, Dzantiev BB (2019) Ways to reach lower detection limits of latera flow immunoassays. https://doi.org/10.5772/intechopen.76926
5. Corstjens P, de Dood CJ, van der Ploeg-van Schip JJ (2011) Lateral flow assay for simultaneous detection of cellular- and humoral immune responses. Clin Biochem 44(14–15):1241–1246
6. Swanson C, D'Andrea A (2013) Lateral flow assay with near-infrared dye for multiplex detection. Clinical Chemistry 59(4 Point-of-care testing):641–648
7. Zhao Y, Wang H, Zhang P et al (2016) Rapid multiplex detection of 10 foodborne pathogens with an up-converting phosphor technology-based 10-channel lateral flow assay. Sci Rep 6:21342. https://doi.org/10.1038/srep21342
8. He PJW, Katis IN, Eason RW, Sones CL (2018) Rapid multiplexed detection on lateral-flow devices using a laser direct-write technique. Biosensors 8:97. https://doi.org/10.3390/bios8040097
9. Carter DJ, Cary RB (2007) Lateral flow microarrays: a novel platform for rapid nucleic acid detection based on miniaturized lateral flow chromatography. Nucleic Acids Res 35(10):e74. https://doi.org/10.1093/nar/gkm269
10. Gantelius J, Bass T, Gundberg A, et al (2011) A ten-minute high density lateral flow protein microarray assay. 15th international conference on miniaturized systems for chemistry and life sciences, 2–6 October 2011, Seattle, WA, USA. 978-0-9798064-4-5/µTAS 2011/$20©11CBMS-0001, pp 1176–1178
11. Taranova NA, Byzova NA, Zaiko VV et al (2013) Integration of lateral flow and microarray technologies for multiplex immunoassay: application to the determination of drugs of abuse. Microchim Acta 180:1165–1172. https://doi.org/10.1007/s00604-013-1043-2
12. Gottfried-Blackmore A, Rubin S, Bai L et al (2020) Effects of processing conditions on stability of immune analytes in human blood. Nat Res Sci Rep 10:2732B. https://doi.org/10.1038/s41598-020-74274-8
13. Matson RS (2021) Chapter 2: Well-based antibody arrays. In: Whittaker KC, Huang R-P (eds) Antibody arrays: methods and protocols, methods in molecular biology, vol 2237. Springer Nature, Cham. https://doi.org/10.1007/978-1-0716-1064-0_2

Chapter 12

A Unique Multiplex ELISA to Profile Growth Factors and Cytokines in Cerebrospinal Fluid

Charitha Madiraju, Amani Sastry, Martha Oppong, Jacob Karp, Maryla Krajewska, Stan Krajewski, Barbara Tomik, Andrzej Szczudlik, and Robert S. Matson

Abstract

Multiplex arrays designed for enzyme-linked immunosorbent assays (ELISAs) are robust and cost-effective for profiling biomarkers. Identification of relevant biomarkers in biological matrices or fluids helps in the understanding of disease pathogenesis. Here, we describe a sandwich ELISA-based multiplex assay to assess growth factor and cytokine levels in cerebrospinal fluid (CSF) samples derived from multiple sclerosis patients, amyotrophic lateral sclerosis patients, and control subjects without any neurological disorder. Results indicate that multiplex assay designed for the sandwich ELISA method is a unique, robust, and cost-effective method for profiling growth factors and cytokines present in CSF samples.

Key words Multiplex array, Sandwich ELISA, Cerebrospinal fluid samples (CSFs), Biomarkers, Growth factors, Cytokines, Multiple sclerosis (MS), Amyotrophic lateral sclerosis (ALS)

1 Introduction

Multiple sclerosis (MS) is a chronic neuroinflammatory demyelinating autoimmune disease of the central nervous system (CNS) [1–3]. Several growth factors and cytokines have been implicated to play a critical role in the disease activation and progression of MS into different clinical subtypes [4–9]. Amyotrophic lateral sclerosis (ALS) is a motor neuron disorder and a neurodegenerative process affecting motor neurons and causing their progressive loss [10, 11]. Like with MS, several growth factors and cytokines play a crucial role toward ALS disease pathogenesis [12–15]. There are diverse views on the paradigm that disease pathogenesis of neurodegenerative diseases like MS and ALS is due to inflammatory cytokines [16–21]. Although it is still unclear in the pathology of

Robert S. Matson (ed.), *ELISA: Methods and Protocols*, Methods in Molecular Biology, vol. 2612,
https://doi.org/10.1007/978-1-0716-2903-1_12,

these diseases, the epidermal growth factor (EGF) and fibroblast growth factor (FGF) levels are known to be altered when assessed in the CSFs from MS and ALS patients [12, 22–25].

Exploring diagnostic application of the disease markers to clinical practice has foreseen various challenges caused by heterogeneity of clinical subtypes of a disease and identification of clinically relevant biomarkers. Assessment of pharmacodynamics with validation of the diagnostic or predictive value of novel biomarkers for the monitoring of therapeutic outcome, disease progression, or remission is calling for robust detection methods [14, 26].

An efficient multiplex profiling platform for establishing biomarker signatures from biological matrices or fluids is germane to this context [12]. Robust and speedy multiplex platforms could test efficiently the clinical biomarkers implicated in the course of disease progression and/or remission. Multiplexing is a very useful tool in research and development (R&D) and clinical diagnostics since it reduces reagent consumption, conserves sample requirements, and reduces labor plus cost per test [27]. The goal of the laboratory protocol described here is to investigate a novel, unique multiplex assay system for the identification of growth factor and cytokine expression applicable to CSFs. Toward this end, a sandwich ELISA-based multiplex assay system for assessing growth factor and cytokine levels in biological matrices and/or body fluids was designed relying on a previously described protocol in a 96-well plate format [27]. Multiplex assay system was demonstrated to have several advantages, including throughput, robustness, uniqueness, simplicity, and cost-efficiency [27, 28].

CSFs derived from MS and ALS patients and age-matched control subjects without neurological abnormalities were investigated in this study. Representative CSFs derived from MS patients diagnosed with either relapsing–remitting MS (RRMS) or primary progressive MS (PPMS) and two unaffected control CSFs derived from the otherwise healthy human subjects are included in the study. Multiplex ELISA analysis included target markers of interest: growth factors EGF and FGF-9, and cytokines IL-6, IFN-γ, IL-4, and IL-1β. Each well in a 96-well plate is imprinted with capture antibodies against the respective target markers, namely, growth factors and cytokines listed above. Sandwich ELISA was performed with either the standard recombinant proteins for the target markers or CSFs (both serving as antigens). The colorimetric read-outs from multiplex assay are analyzed using ImageJ. The materials needed for multiplex assay system, methodology for multiplex assay system, results, and protocol notes are described below. Preliminary results identify the multiplex assay to be a unique and cost-effective method for establishing the expression profile of target markers present in CSFs.

2 Materials

Multiplex ELISA for standards and samples

1. Growth factor and cytokine multiplex array 96-well test plate (*see* **Note 1**).
2. Wash buffer, Tris-buffered saline with 0.1% Tween-20® detergent (TBST) (1X), pH 7.4.
3. Blocking buffer, 1% BSA (w/v) in 1X phosphate-buffered saline (PBS), pH 7.4.
4. Sample buffer, 1X PBS pH 7.4.
5. Standards: Recombinant growth factors and cytokines (*see* **Note 2**).
6. Biotinylated detection antibodies (*see* **Note 3**).
7. Test samples: CSFs were derived from MS/ALS patients and controls without neurological symptoms (*see* **Note 2**).
8. Streptavidin-horseradish peroxidase (S-HRP) enzyme conjugate.
9. 3,3′, 5,5′ Tetramethylbenzidine (TMB) blotting – precipitating chromogenic enzyme substrate.
10. Eppendorf tubes: 500 μL to 1.5 mL.
11. Single-channel and multi-channel pipettes.
12. Pipette tips: 1–10 μL, 20–200 μL; 200–1000 μL delivery range.
13. Buffer reservoir.
14. Charge-coupled device (CCD) camera colorimetric reader or digital imaging-type microscope.

3 Methods

3.1 Multiplex ELISA Protocol for Standards and Samples

1. Ascertain wash buffer, sample buffer, and blocking buffer are ready to use before the start of the assay (*see* **Note 4**).
2. Remove the assay strip out of the imprinted multiplex 96-well array plate and bring to room temperature (RT) 30 min before performing the experimental protocol (*see* **Note 5**).
3. Pipette 100 μL 1X TBST pH 7.4 wash buffer into assay wells and incubate at RT for 5 min (*see* **Note 6**).
4. Rinse the assay wells once to remove the wash buffer (*see* **Note 7**).
5. Pipette 100 μL of 1% (w/v) BSA in 1X PBS pH 7.4 blocking buffer into the assay wells.

6. Incubate the assay strip with blocking buffer for 1 hr at RT (*see* **Note 8**).
7. Prepare a working stock mix of known concentrations of the recombinant growth factors and cytokines that serve as standards in 1X PBS pH 7.4 sample buffer (*see* **Notes 9** and **10**).
8. Perform 1:2 serial dilution of the working stock mix of the standards from **step 7** (*see* **Note 11**).
9. Rinse the blocked assay wells three times, each rinse with 100 μL 1X TBST pH 7.4 wash buffer.
10. Pipette 100 μL each of the serially diluted standards from **step 8** into the assay wells. Include a zero-concentration sample of 100 μL of 1X PBS 7.4 sample buffer.
11. Prepare 100 μL of CSFs derived from MS or ALS patients and controls:
    (a) Thaw CSFs on ice.
    (b) Dilute samples, if needed, in 1X PBS pH 7.4 sample buffer.
    (c) Deliver 100 μL of each sample into designated wells.
12. Incubate the assay strip with the standards and samples for 1 h at RT.
13. Prepare a mix of biotin-labeled detection antibodies each at 1: 1000 (v/v) dilution in 1% (w/v) BSA in 1X PBS pH 7.4 blocking buffer (*see* **Note 12**).
14. Rinse the assay strip incubated with standards and samples in **step 12** three times, each rinse with 100 μL 1X TBST pH 7.4 wash buffer.
15. Add 100 μL of biotin-labeled detection antibody mix from **step 13** into each well of the assay strip.
16. Incubate the assay strip with biotinylated antibodies for 1 h at RT.
17. Prepare S-HRP enzyme conjugate at 1:200 (v/v) dilution in 1% (w/v) BSA in 1X PBS pH 7.4 blocking buffer (*see* **Note 13**).
18. Rinse the plate three times, each rinse with 100 μL 1X TBST pH 7.4 wash buffer.
19. Add 100 μL of S-HRP reagent at 1:200 (v/v) dilution into each well.
20. Incubate the assay strip with S-HRP for 20 min at RT.
21. Rinse the assay well three times, each rinse with 100 μL 1X TBST pH 7.4 wash buffer.
22. Add 100 μL of the chromogenic TMB precipitating blotting reagent to each assay well.

23. Incubate the assay wells with the TMB precipitating blotting solution for 15 min at RT in the dark for color development (*see* **Note 14**).
24. Remove the TMB solution but do not rinse (*see* **Note 15**).
25. Observe the precipitated blots in wells using either a colorimetric reader or a digital microscope and capture the images for further analysis.
26. Perform ImageJ analysis of the blots to quantify the data with the standards first to build standard curves for the respective growth factors and cytokines (*see* **Note 16**).

3.2 Results

The multiplex sandwich ELISA plate is designed to assess growth factors and cytokines in biological matrices or body fluids. A schematic representation of multiplex sandwich ELISA is shown in Fig. 1. Capture antibodies against target markers, namely, growth factors (EGF and FGF-9) and cytokines (IFN-γ, IL-6, IL-1β, and IL-4) were imprinted into each well of a 96-well plate in a multiplex array format. A schematic representation of the imprinted

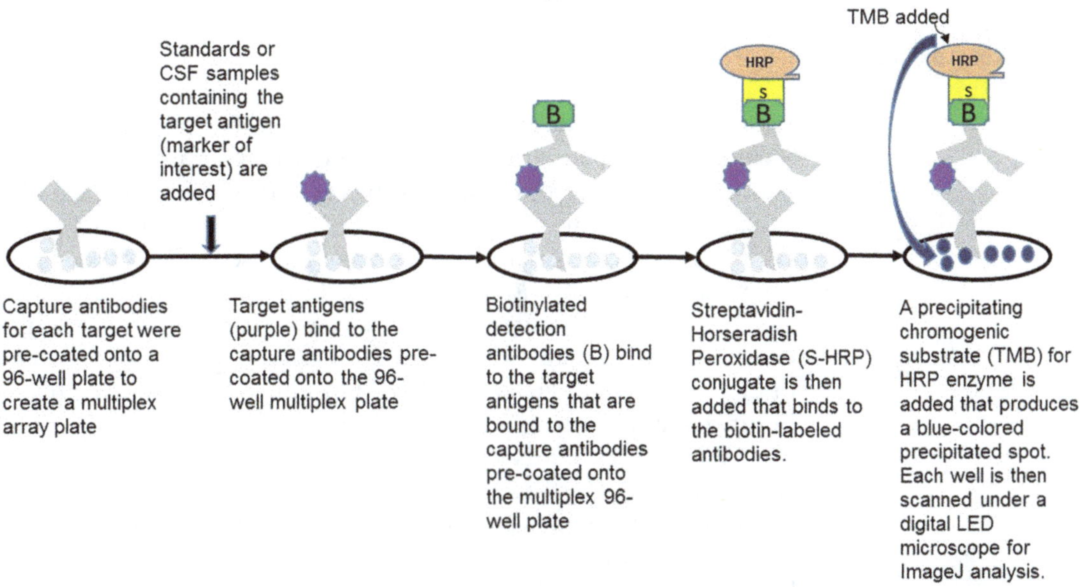

Fig. 1 Schematic representation of multiplex ELISA. A 96-well microtiter plate is precoated with target-specific capture antibodies (against IL-6, EGF, FGF-9, IL-1β, IL-4, and IFN-γ). Standards or samples are added, and the target antigen present in the standards or samples binds to the capture antibody. A biotin-conjugated detection antibody is then added, which binds to the captured antigen. A streptavidin-horseradish peroxidase (S-HRP) conjugate is then added that binds to the biotin. A chromogenic (TMB) substrate is then added that reacts with the HRP-derived product, resulting in a blue-colored precipitate. Each well is scanned using a digital LED microscope. Pixel intensities of the precipitated spots are measured using ImageJ. Standard graphs are generated for each marker, unknowns determined, and data quantified and presented as mean ± SD

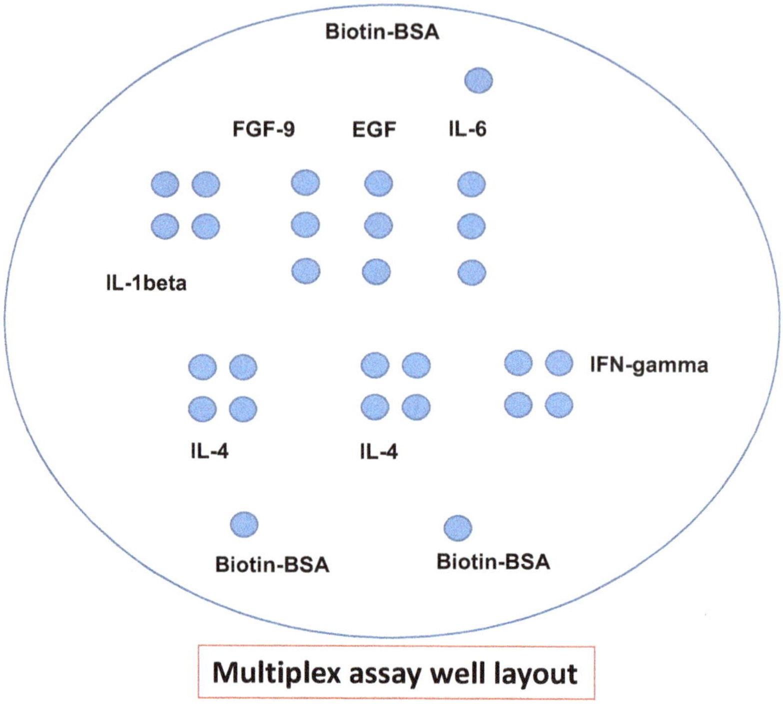

Fig. 2 Schematic representation of multiplex ELISA plate layout

multiplex array plate is shown in Fig. 2. Sandwich ELISA is performed using recombinant protein standards against target markers and CSFs derived from control subjects without neurological disease, relapsing–remitting MS (RRMS), primary progressive MS (PPMS), and ALS patients. A schematic representation of the multiplex assay methodology is shown in Fig. 3.

The entire immunoassay is conducted under static (no shaking) conditions except for the rinsing or washing steps. Growth factors or cytokines present in standard and test samples are detected using biotinylated anti-growth factor or anti-cytokine detection antibodies. S-HRP that binds to the biotinylated detection antibodies is added followed by addition of TMB precipitating chromogenic substrate that undergoes enzymatic hydrolysis in the presence of S-HRP, which then produces dark blue colored spots where analytes are present as shown in the schematic representations in Figs. 1 and 2. Multiplex assay protocol methodology is shown in Fig. 3. Multiplex assay results depicted in three representative assay strips shown in Fig. 4 identified serially diluted known concentrations of recombinant protein standards (shown as dark spots for the respective target markers), binding to the capture antibodies imprinted into each well of the assay strip. Representative images from two standards and two samples are also shown.

Multiplex assay wells were experimented for any cross-reactivity of the imprinted captured antibodies, and this was done by assessing only one target growth factor or cytokine in each well. Results

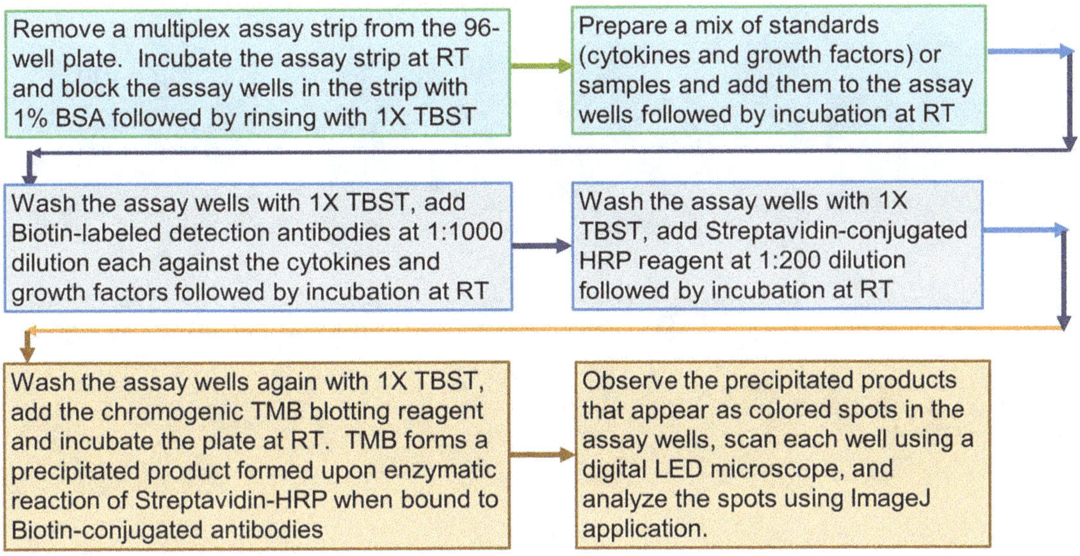

Fig. 3 Schematic representation of the multiplex ELISA protocol methodology

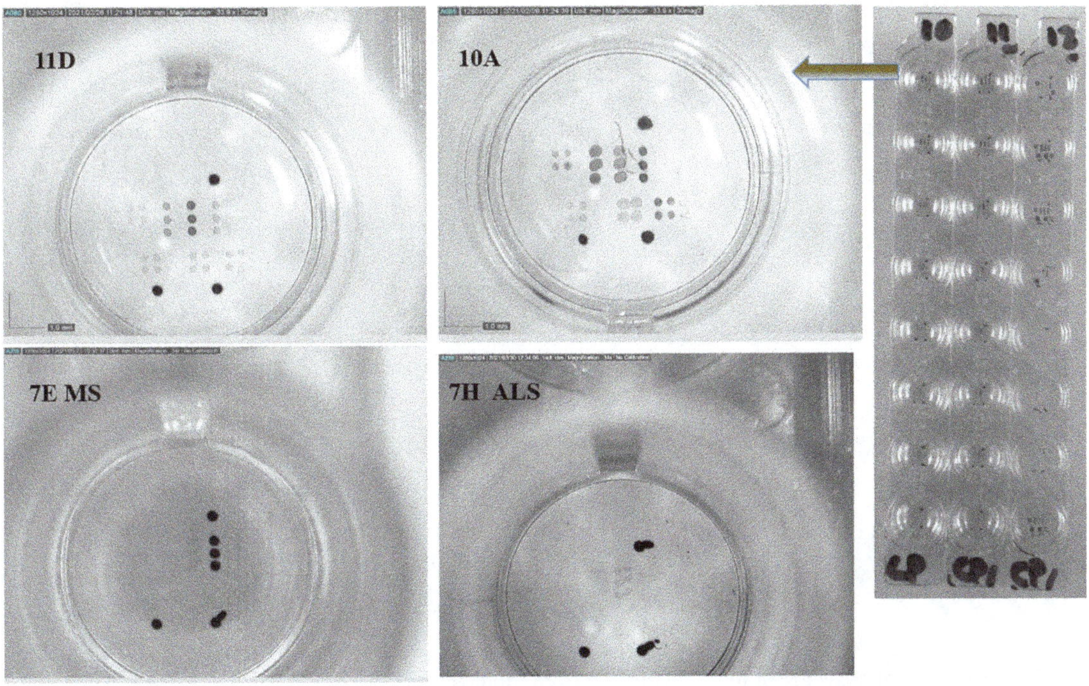

Fig. 4 Representative multiplex ELISA strips upon image analysis, including images from two serially diluted standards (labeled 11D, 10A) and two samples-MS (labeled 7E), ALS (labeled 7H), are shown

presented in Fig. 5 show no cross-reactivity observed when one recombinant protein standard for a target growth factor or cytokine was tested alone in each well. This indicates a pooled mixture of standards for the target growth factors and cytokines investigated in this study protocol could be used when doing multiplex assay.

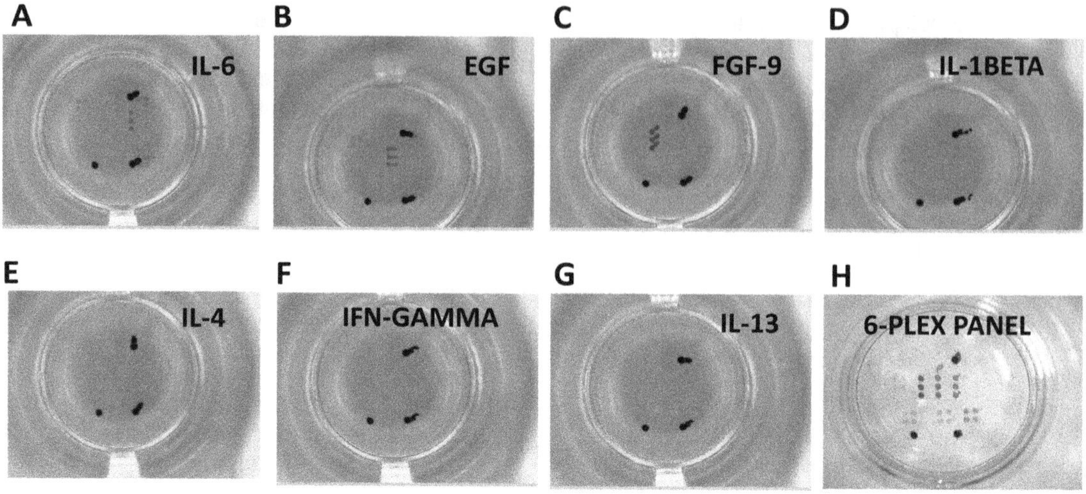

Fig. 5 (**a**–**h**) Cross-reactivity testing of the multiplex ELISA plate
The cross-reactivity was assessed for only one target cytokine or growth factor in each well. The inset of each of the figures has the respective cytokines or growth factors labeled. The inset of 5H has the label for the 6-Plex panel used in the multiplex ELISA

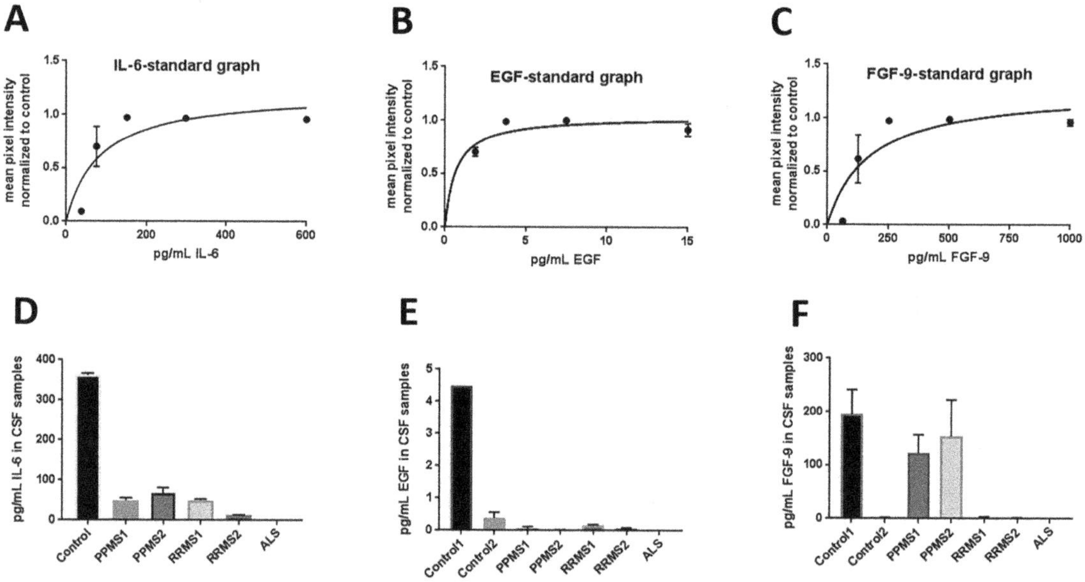

Fig. 6 (**a**–**f**) Graphical analyses of the data derived from multiplex ELISA
Standard graphs for target markers IL-6, EGF, and FGF-9 (**a**–**c**) and data for CSFs derived from controls and MS or ALS patients (**d**–**f**) are presented as mean ± SD. Normalized mean pixel intensity values are presented

Serially diluted standards for target markers and CSFs were then used to assess the suitability of this unique multiplex system in detecting target markers (Fig. 6a–c). Standard graphs for IL-6, EGF, and FGF-9 are shown in Fig. 6a–c, respectively.

Graphical analyses of the interpolated data derived from multiplex analysis of IL-6, EGF, and FGF-9 in CSFs (derived from controls, MS, and ALS patients) are presented as mean ± SD in Fig. 6d–f, respectively. Normalized mean pixel intensity values are presented. Detectable levels of serially diluted IL-6, EGF, and FGF-9 are present in the standard samples. Analysis of CSFs from MS and ALS patients suggests that this format of multiplex assay detected lower levels of EGF, FGF-9, and IL-6 in patient CSF samples compared to that observed in CSFs from controls (Fig. 6d–f). Detectable levels of IL-6, EGF, and FGF-9 were present in CSFs from MS patients. Since some of the ALS CSFs' lyophilizates were protein rich, which required dilution and gentle sonication to be resolved, it may have influenced the levels of the target markers. However, we did not concentrate them in a speed-vac, which might be applied to help increase the detection sensitivity of targeted biomarkers. Although there is a difference in the levels of these markers in patient CSFs versus controls without neurological symptoms, at this point, the goal is not to observe the differences in expression profile of the markers between two different subsets. Future studies will focus on establishing an expression profile of target markers in comparison with control subjects.

4 Notes

1. The detailed protocol of the imprinting procedure for creating a multiplex antibody array plate for sandwich ELISA is described elsewhere [27]. The multiplex panel used for this study is comprised of purified anti-growth factor (EGF and FGF-9) and anti-cytokine antibodies (IL-1β, IL-4, IL-6, IFN-γ) arrayed at the bottom of the strip-well microtiter 96-well plate. Imprinted multiplex microarray test plate is placed with a desiccant in a sealed pouch and stored at 2–8° C. Do not freeze the plate. Recommended plates: Scienion sciPLEXPLATE 96, Type 2 or Greiner Bio, High Binding.
2. Standards included human IL-4 standard (BioLegend 79022); human IL-13 standard (BioLegend 78298); human IFN-γ standard (BioLegend 79103); human IL-6 standard (BioLegend 570809); human IL-1β standard (BioLegend 78164); human EGF standard (R&D Systems 841295); human FGF-9 standard (R&D Systems 840906). CSFs from MS patients were obtained from the Human Brain and Spinal Fluid Resource Center, VA Greater Los Angeles Healthcare System. All CSFs from ALS patients were lyophilized prior to shipping. They were reconstituted to 1 mL of original volume prior to

testing. If any problem is encountered while resolving the precipitates, dilution and gentle sonication was applied. Aliquoted CSFs were frozen and stored at −80° C.

3. Complementary detection antibodies matched to capture antibodies used with array plate: biotinylated anti-EGF, biotinylated anti-FGF-9, biotinylated anti-IL-6, biotinylated anti-IL-1β, biotinylated anti-IL-4, and biotinylated anti-IFN-γ.
4. Assure the sample buffer, wash buffer, and blocking buffer are prepared and are ready to use. Sample buffer, wash buffer, and blocking buffer used in the assay should be clean and clear and should not contain visible particulate matter.
5. Carefully remove the imprinted 96-well multiplex plate from the storage pouch, select the assay strip or strips for use, and remove the assay strip or strips out of the 96-well plate 30 min prior to starting the assay. Set the assay strip at RT before use. Return remaining plate back to the pouch and seal for storage at 4 °C until later use. Do not freeze the imprinted multiplex plate at any point.
6. Carefully pipette reagents into assay wells. Do not touch pipette tip to the well surface.
7. During rinsing or washing steps, assure removal of the liquid by rapid and gentle inversion of the strip into a waste container or sink. Do not let wells dry out. Blot strip using lint-free towels after the final rinse step.
8. Perform incubation steps at ambient RT (22–25 °C) without mixing or shaking.
9. Gently mix and avoid foaming while preparing the samples. Use freshly prepared standards or test samples.
10. Standards were tested in 50 mM HEPES, pH 7.4 but did not show much difference in assay results compared to when prepared in 1X PBS, 7.4.
11. Use 1.5-mL Eppendorf tubes or a clear 96-well plate to prepare serially diluted concentrations of standards for growth factors and cytokines.
12. The mix should contain the respective growth factors and cytokines present in the assay well layout (as shown in Fig. 2). Prepare fresh biotinylated detection antibody stock mix, each detection antibody at 1:1000 (v/v) in blocking buffer each time the assay is performed.
13. Prepare fresh S-HRP stock 1:200 (v/v) in blocking buffer each time the assay is performed. Recommended conjugate, S-HRP 890803 (R&D Systems).
14. Wrap the aliquot of TMB solution with aluminum foil to avoid exposure to natural light. All incubations with TMB solution

should be done in the dark. Bring TMB solution to RT prior to use. Do not discard solution after use. Place it back in the refrigerator for future use. Recommended reagent: Thermo Scientific, 1-Step Ultra TMB-Blotting Solution (ready to use).

15. To avoid distortion or reflection from TMB liquid during imaging, it is advisable to remove the solution in advance. This is accomplished by blotting the inverted strip onto paper wipes.
16. Keep the threshold values the same for all the standards and samples when doing ImageJ analysis. GraphPad Prism v7 may be used for graphing the standard graphs.

Acknowledgments

In memoriam: Submitted as a tribute to honor **Dr. Barbara Tomik** for her valuable contributions toward the clinical data collection and preparation of patient samples.

References

1. Noseworthy JH, Lucchinetti C, Rodriguez M et al (2000) Multiple sclerosis. N Engl J Med 343(13):938–952
2. Reich DS, Lucchinetti CF, Calabresi PA (2018) Multiple sclerosis. N Engl J Med 378(2):169–180
3. Sospedra M, Martin R (2005) Immunology of multiple sclerosis. Annu Rev Immunol 23: 683–747
4. Göbel K, Ruck T, Meuth SG (2018) Cytokine signaling in multiple sclerosis: lost in translation. Mult Scler 24(4):432–439
5. Huang Y, Dreyfus CF (2016) The role of growth factors as a therapeutic approach to demyelinating disease. Exp Neurol 283 (Pt B):531–540
6. Imitola J, Chitnis T, Khoury SJ (2005) Cytokines in multiple sclerosis: from bench to bedside. Pharmacol Ther 106(2):163–177
7. Klineova S, Lublin FD (2018) Clinical course of multiple sclerosis. Cold Spring Harb Perspect Med 8(9):a028928
8. Maimone D, Gregory S, Arnason BG et al (1991) Cytokine levels in the cerebrospinal fluid and serum of patients with multiple sclerosis. J Neuroimmunol 32(1):67–74
9. Matsushita T, Tateishi T, Isobe N et al (2013) Characteristic cerebrospinal fluid cytokine/chemokine profiles in neuromyelitis optica, relapsing remitting or primary progressive multiple sclerosis. PLoS One 8(4):e61835
10. Hardiman O, Al-Chalabi A, Chio A et al (2017) Amyotrophic lateral sclerosis. Nat Rev Dis Primers 3:17071
11. Lyon MS, Wosiski-Kuhn M, Gillespie R et al (2019) Inflammation, immunity, and amyotrophic lateral sclerosis: I. Etiology and pathology. Muscle Nerve 59(1):10–22
12. Cao MC, Cawston EE, Chen G et al (2022) Serum biomarkers of neuroinflammation and blood-brain barrier leakage in amyotrophic lateral sclerosis. BMC Neurol 22(1):216
13. Huang F, Zhu Y, Hsiao-Nakamoto J et al (2020) Longitudinal biomarkers in amyotrophic lateral sclerosis. Ann Clin Transl Neurol 7(7):1103–1116
14. Staats KA, Borchelt DR, Tansey MG et al (2022) Blood-based biomarkers of inflammation in amyotrophic lateral sclerosis. Mol Neurodegener 17(1):11
15. Webster HD (1997) Growth factors and myelin regeneration in multiple sclerosis. Mult Scler 3(2):113–120
16. Martino G, Furlan R, Brambilla E et al (2000) Cytokines and immunity in multiple sclerosis: the dual signal hypothesis. J Neuroimmunol 109(1):3–9
17. Moreau C, Devos D, Brunaud-Danel V et al (2005) Elevated IL-6 and TNF-alpha levels in patients with ALS: inflammation or hypoxia? Neurology 65(12):1958–1960

18. Petković F, Castellano B (2016) The role of interleukin-6 in central nervous system demyelination. Neural Regen Res 11(12):1922–1923
19. Sharief MK (1998) Cytokines in multiple sclerosis: pro-inflammation or pro-remyelination? Mult Scler 4(3):169–173
20. Wagner CA, Roqué PJ, Goverman JM (2020) Pathogenic T cell cytokines in multiple sclerosis. J Exp Med 217(1):e20190460
21. Wullschleger A, Kapina V, Molnarfi N et al (2013) Cerebrospinal fluid Interleukin-6 in central nervous system inflammatory diseases. PLoS One 8(8):e72399
22. Cieślak D, Szulc-Kuberska J, Stepień H et al (1986) Epidermal growth factor in human cerebrospinal fluid: reduced levels in amyotrophic lateral sclerosis. J Neurol 233(6): 376–377
23. Kanda T, Iwasaki T, Nakamura S et al (1999) FGF-9 is an autocrine/paracrine neurotrophic substance for spinal motoneurons. Int J Dev Neurosci 17(3):191–200
24. Scalabrino G (2022) Newly identified deficiencies in the multiple sclerosis central nervous system and their impact on the remyelination failure. Biomedicine 10(4):815
25. Scalabrino G (2022) Epidermal growth factor in the CNS: a beguiling journey from integrated cell biology to multiple sclerosis. An extensive translational overview. Cell Mol Neurobiol 42(4):891–916
26. Paul A, Comabella M, Gandhi R (2019) Biomarkers in multiple sclerosis. Cold Spring Harb Perspect Med 9(3):a029058
27. Matson RS (2021) Well-based antibody arrays. Methods Mol Biol 2237:11–38
28. Matson RS (2015) Multiplex ELISA using oligonucleotide tethered antibodies. Methods Mol Biol 1318:181–195

Chapter 13

Well-Based Multiplex Food Allergen Colorimetric ELISA

Robert S. Matson

Abstract

Food allergy is a well-recognized and significant health hazard around the world. At least 160 food groups have been identified that present allergenic reactions or other sensitivities and intolerance in humans. Enzyme-linked immunosorbent assay (ELISA) is an accepted platform for identifying the nature of the food allergy and its severity. It is now possible to simultaneously screen patients for allergic sensitivity and intolerance to multiple allergens using multiplex immunoassays. This chapter describes the preparation and utility of a multiplex allergen ELISA for the assessment of food allergy and sensitivity in patients.

Key words Multiplex, Colorimetric, HRP, ELISA, Immunoassay, Strip-well, Microplate, Microarray, Allergens, Protein extracts, Plasma, FDA, Allergy, Intolerance, IgE, IgG, IgG4

1 Introduction

Food allergies represent an increasingly significant health issue around the world. More than 160 foods are known to cause allergic reactions in humans. The U.S. Food and Drug Administration (FDA) has identified eight food groups that are responsible for 90% of clinically defined (IgE-mediated) acute food allergies [1, 2]. *The eight foods identified by the law are*

1. Milk
2. Eggs
3. Fish (e.g., bass, flounder, cod)
4. Crustacean shellfish (e.g., crab, lobster, shrimp)
5. Tree nuts (e.g., almonds, walnuts, pecans)
6. Peanuts
7. Wheat
8. Soybeans

Robert S. Matson (ed.), *ELISA: Methods and Protocols*, Methods in Molecular Biology, vol. 2612,
https://doi.org/10.1007/978-1-0716-2903-1_13,

Table 1
Comparison of food allergy and food intolerance

Food allergy	Food intolerance
A sudden change (acute condition) "Triggered" by ingestion of small amounts of the food Occurs each time you eat the food Potentially life-threatening	Gradual change (chronic condition) Can occur after consuming large portions or from the frequent ingestion of the same food Not life-threatening
Common symptoms Rash, hives, or itchy feeling Shortness of breath Difficulty in swallowing Chest pain Anaphylactic shock	Common symptoms Gas, cramps, or bloating Heartburn Headaches Irritable or nervous condition
Example: peanut allergy	Examples: wheat gluten (non-celiac gluten sensitivity -NCGS) Irritable bowel syndrome (IBS) Milk (lactate) intolerance is Enzyme deficiency—similar symptoms but not an allergic reaction
Associated with increased levels of IgE antibodies as determined by ELISA	Associated with increased levels in blood of IgG antibodies specific to the food. In particular, the IgG4 sub-class is often assessed by ELISA as an indicator of food intolerance
Clinically defined	No clinical test
IgE	IgG (IgG4)

The FDA enforces the Food Allergen Labeling and Consumer Protection Act of 2004, which requires the labeling of food products as to the source of the ingredients that may contain these regulated food groups. Moreover, less-well-understood chronic conditions with symptoms such as cramps, bloating, heartburn, headaches, or other irritabilities, commonly referred to as food intolerance, are studied here (*see* Table 1).

We describe the steps in preparing a multiplex, colorimetric enzyme-linked immunosorbent assay (ELISA) (indirect) for the assessment of food allergy and intolerance in patients (Fig. 1). The developed assay is based upon the immobilization of allergenic food protein extracts for diagnostic purpose [3] in microarray format such as previously accomplished on glass microscope or nitrocellulose slides [4–7], however, in this case contained in microwell plates [8]. Food-specific antibodies present in the patient's plasma bind to the corresponding specific food protein allergen. An HRP conjugate antihuman antibody is used for the detection of a TMB-insoluble product that produces a dark-blue color. The resulting in-well microarray is scanned, and the images analyzed.

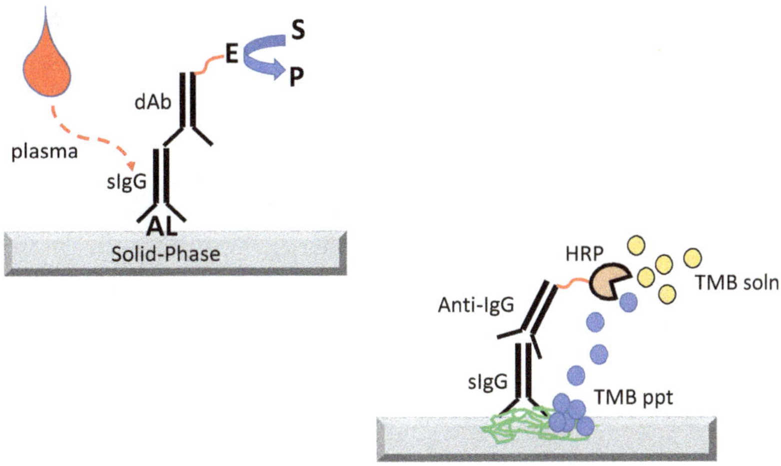

Fig. 1 Colorimetric ELISA for allergen. Allergen (AL) protein extract is directly adsorbed to the solid phase. Plasma containing the antibody specific to the allergen (sIgG) is separated from blood and applied to the solid phase. Next, a detection antibody (dAb) that is anti-IgG binds to (sIgG). HRP enzyme (E) tethered to the dAb catalyzes the conversion of TMB substrate (S) to a signal product (P), which precipitate at the site, thereby confirming the presence of the allergenic sIgG

The chapter is divided into three main sections: preparation of food protein extracts, spotting of the extracts by microarraying into wells, and development of the multiplex food allergen colorimetric ELISA.

2 Materials

2.1 Extraction of Allergens

1. Food produce (*see* **Note 1**).
2. Kitchen utensils: Chopping board, knives, blender, strainer.
3. Top loading, tare weight scale: 300-g capacity.
4. Centrifuge: Swinging bucket type for 50-mL centrifuge tubes, 3000–4000 rpm.
5. Centrifuge: Fixed angle, high speed for 2-mL centrifuge tubes, 10,000 rpm.
6. Magnetic stir plate, stir bars.
7. Distilled water.
8. Ammonium sulfate.
9. Dialysis tubing or device (*see* **Note 2**).
10. PBS, pH 7.4, sterile-filtered.
11. Micro BCA Protein Assay Kit (Thermo Scientific, Cat# 23235).

2.2 Printing of Allergens

1. Solid phase: Polystyrene-based microtiter plates for binding proteins (*see* **Note 3**).
2. Allergens: Food protein extracts (*see* **Note 4**).
3. Controls: Human IgG, human IgG4, human IgE, natural or recombinant, purified.
4. Print buffer: Phosphate-buffered saline, sterile-filtered (PBS, Sigma-Aldrich).
5. Source plate: Nonbinding, 384-well polystyrene or polypropylene plate.
6. Microarray dispenser: Capable of droplet dispense from ~1 to 10 nL (*see* **Note 5**).
7. Microcentrifuge: 0–2000 rpm.
8. Rotator: End-over-end type or nutating mixer.
9. Pipettors: Single-channel, multichannel, positive displacement type.
10. Pipette tips: 1–10 μL, 20–200 μL, 200–1000 μL capacity.

2.3 Colorimetric ELISA

1. Assay plate: Arrayed food extracts adsorbed to polystyrene wells (*see* **Note 6**).
2. Wash buffer: TBST (1×), pH 7.4.
3. Blocking buffer: 1% BSA (V)–PBS, pH 7.4 (*see* **Note 7**).
4. Enzyme conjugate: HRP antihuman antibody (*see* **Note 8**).
5. Enzyme substrate: TMB blot precipitating substrate (*see* **Note 9**).
6. Plasma samples (*see* **Note 10**).
7. Microplate or strip-well reader: Colorimetric, imaging type (*see* **Note 11**).

3 Methods

3.1 Extraction of Allergens

1. Weigh out ~50–100 g of raw food material for extraction (*see* **Note 12**).
2. Rinse in DDI water and then blot with paper towels.
3. Chop or otherwise section (dice) into workable portions with a knife.
4. Place diced food in a glass beaker and add ice-cold PBS buffer.
5. Homogenize using a handheld blender (*see* **Note 13**).
6. Strain homogenate over a beaker using stainless steel strainer to separate the solution from the solids (paste). Press pastes with a spatula to remove liquid. Discard paste.

7. Centrifuge strained homogenate in 50-mL tubes at 3000 rpm, 10 min.
8. Decant solution from the tube into a new beaker. Repeat centrifugation of any additional homogenate.
9. Centrifuge decants and discard pellet. Save the supernatant solution (crude extract) in a beaker and place on ice.
10. Ammonium sulfate (AS) precipitation:
    (a) Determine the volume of the decanted (crude extract) solutions.
    (b) Transfer combined decants to a beaker and stir on ice.
    (c) Determine the amount of AS required to achieve 75% saturation, 4 °C (*see* **Note 14**).
    (d) Slowly add AS over 30 min or longer with stirring.
    (e) Transfer to 50-mL centrifuge tubes.
    (f) Allow AS solution to settle overnight, 4 °C.
    (g) Centrifuge (50-mL tubes) at 4000 rpm, 10 min.
    (h) Decant off supernatant (AS supt.) using a pipette.
    (i) Combine the AS precipitates (AS ppt. pellets) from the tubes. This is the source of the precipitated protein from the extract.
11. Centrifuge at 4000 rpm, 10 min, AS ppt. to collect pellet (precipitated proteins).
12. Transfer to 2-mL tubes and pellet at 10,000 rpm, 10 min.
13. Weigh pelleted AS fractions (AS ppt.) to be dialyzed.
14. Dialyze AS ppt. Against PBS buffer (*see* **Note 15**).
15. Determine protein content using the micro-BCA reagent.
16. Store at 4 °C for short term.

3.2 Printing of Allergens

The Food Allergen Panel is comprised of food extracts arrayed in the bottom of microtiter plate wells. In this version, the extracts are prepared in print buffer and dispensed to the surface of the wells of strip-well plates where they are physically adsorbed. The plate is allowed to dry and then subsequently blocked in 1% BSA.

Protocol

It is highly recommended that the user wear gloves during these procedures to avoid contamination issues and accumulation of debris on surfaces.

1. Prepare the probes (allergen extracts, controls) for printing in PBS buffer (*see* **Note 16**).
2. Place 70 μL of each probe print solution in an assigned well of a 384-well source plate.

3. Lid the plate (*see* **Note 17**).
4. Spin the plate at 2000 rpm, 5 minutes to remove air.
5. Set up the microarray dispenser for printing. Critical conditions are relative humidity (RH), temperature, and dispense element cleaning. Set RH ~ 60%, temperature ~25 °C (*see* **Note 18**).
6. Place the source plate on the deck of the microarray dispenser.
7. Carefully position a clean, flat-bottom, polystyrene plate to be used for creation of the well arrays on the deck.
8. Conduct the print run cycles.
9. Allow the printed droplets on the plate to completely dry down in a clean environment (*see* **Note 19**).

3.3 Colorimetric ELISA

Diluted plasma thought to contain antibodies generated to allergenic or food intolerance response are applied. Except for the washing steps, the entire immunoassay is conducted under static (no shake) conditions. Food allergen-specific antibodies (sIgE/sIgG/sIgG4) are detected using HRP antihuman IgG/IgG4 or IgE conjugates. A TMB precipitating substrate is applied that produces a dark-blue color in the presence of HRP (Fig. 1).

Protocol

1. Allow the plate to reach room temperature prior to use. Carefully remove the strip-well plate from the storage pouch and select the strips to be used. Return the rest of the plate back to its pouch and seal for storage.
2. Deliver 100 μL of wash buffer (1×) to each well to be used. Do not touch the well surface with the pipette tip.
3. Let soak for 5 min and then remove liquid by rapid inversion of the strip over a sink or waste container. Blot the strip using lint-free towels.
4. Prepare samples: Dilute plasma in blocking buffer and gently mix. Avoid foaming (*see* **Note 20**).
5. Deliver 100 μL to each well and set aside. Incubate 1 h at ambient temperature (22–25 °C) without mixing or shaking (*see* **Note 21**).
6. Remove liquid by rapid inversion of the strip over a sink or waste container.
7. Deliver 100 μL wash buffer (1×) to each well. Remove solution. Repeat rinse two times, and then let soak 5 min. Following the soak, rinse an additional three times. Do not allow the wells to dry out. After final rinse, blot the strip using lint-free towels.

8. Prepare fresh enzyme conjugate(s): Dilute stock 1:1000 (v/v) in blocking buffer (*see* **Note 22**).
9. Deliver 100 μL to each conjugate (HRP antihuman IgG/IgG4 or HRP antihuman IgE) to separate wells and set aside. Incubate 30 min. No mixing or shaking.
10. Remove liquid and rinse as previously described (*see* **step** 7).
11. Prepare substrate: Bring TMB solution to room temperature prior to use. Cover reagent bottle or tube to avoid excessive light exposure. Deliver 100 μL to each well and set aside. Incubate 15 min. No mixing or shaking. Do not discard solution (*see* **Note 23**).
12. Read the strip (*see* **Note 24**).

Example

A patient of known allergy to dairy products was evaluated for food sensitivity using the well-based multiplex allergen ELISA. Each well of a 96 strip-well plate (sciPLATE Type 2, Scienion AG) was arrayed with 90 food protein extracts along with serial dilutions of human IgG that served as internal controls, as well as for construction of the dose–response curve for estimation of food-specific allergen IgG (total) response. The plate was printed using a sciFLEXARRAYER S5 equipped with a single piezo capillary dispenser (PCD). The runtime for dispensing using a single PCD was approximately 10 hours. The printed plate was processed according to Protocol in Subheading 3.2 and the ELISA developed for total plasma IgG as provided in Protocol in Subheading 3.3 (Fig. 2). Subsequently, the pixel intensity values for each spot in the array were obtained from plate well imaging scans using the sciREADER CL2. The resulting .csv file was exported into EXCEL and the specific IgG dose associated with each food allergen determined by interpolation based upon linear regression analysis of the in-well IgG control spot values (Fig. 3).

4 Notes

1. *Food produce.* Select raw foods, uncooked or unprocessed, if possible, to avoid potential interferences due to food additives such as preservatives.
2. *Dialysis.* Transfer 400–600 μL protein solution to Slide-A-Lyzer cassettes (Pierce # 66333; 3.5 K MWCO) or equivalent product. Dialysis tubing can also be used for this purpose.
3. *Protein-binding plates.* It is critical to use a medium- to high-protein-binding polystyrene plate designed specifically for use

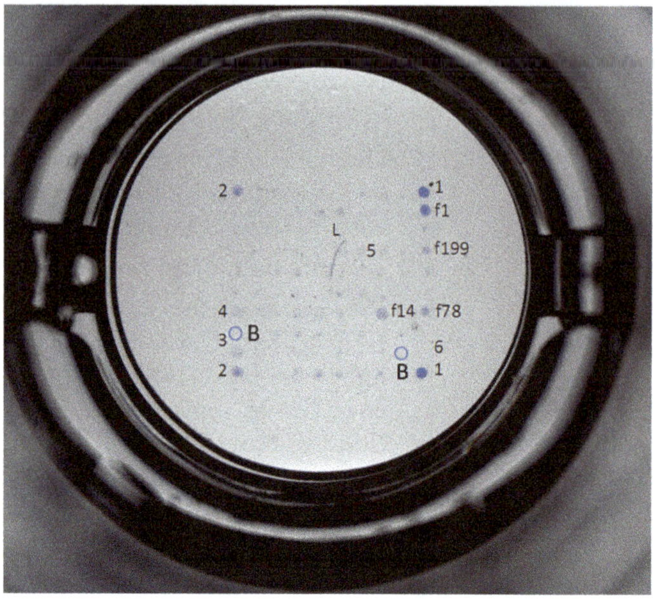

Fig. 2 Multiplex allergen colorimetric ELISA of patient plasma sample. A 10 × 10 array of food protein extracts ($n = 90$), IgG controls ($n = 8$), and buffer blanks ($n = 2$) were printed in each well of a 96-strip well plate. A 1:100 (v/v) dilution of plasma was applied to the well and processed according to Protocol 3.3. Signal for specific IgG was generated using HRP antihuman IgG with 1-Step Ultra TMB-Blotting Solution. Predominant signals were obtained for egg white (f1), cow's milk (f199), casein (f78), and soybean (f14). Human IgG controls (μg/mL): (1) 125, (2) 62.5, (3) 31.3, (4) 15.6, (5) 7.8 and (6) 3.9; L = lint; B = blank. Image analysis was done using the sciREADER CL2 (Scienion AG) (*see* **Note 25**)

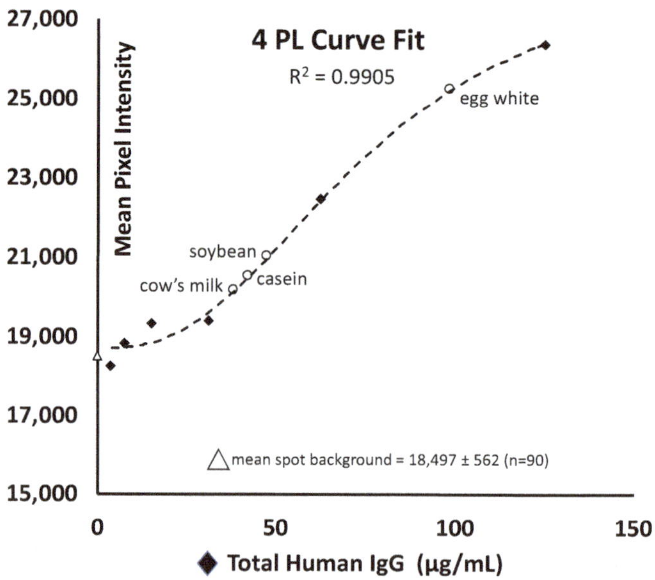

Fig. 3 Determination of food allergen-specific IgG response. Using the IgG control spot pixel intensities, a separate standard curve is generated for each well. The concentration of sIgG associated with each spot (food extract) is interpolated from the pixel intensity using the 4 PL regression (AssayFit Pro 1.31) (*see* **Note 26**)

in ELISA applications. For example, Greiner Bio-One, High Bind plate or Scienion, sciPLEXPLATE 96, Type 2.

4. *Allergens.* In this application, the extracted proteins from food sources are the source of the allergen. Most represent crude, unpurified mixtures of proteins. The quantity of putative allergen(s) is largely unknown. For this reason, adjustment of the total protein concentration used in printing may be necessary to achieve adequate dose–response.

5. *Microarray dispenser.* For printing of the extracts onto polystyrene surfaces, the dispense droplet volume is about 1–2 nL per spot. This requires a low-volume dispenser system (microarrayer) capable of high-precision dispensing. For example, deposit of 100-micron diameter spots on a 300-μm pitch within a well. Noncontact piezo dispensing is recommended such as available from the following vendors: Scienion AG, Berlin, Germany (www.scienion.com); BioDot, Irvine, CA (www.biodot.com); ArrayJet, UK (www.arrayjet.co.uk); GeSiM mbH, Germany (www.gesim.de); and M2-Automation, Berlin, Germany (www.m2-automation.com).

6. *Food allergen assay plate.* Previously blocked, stored dry in sealed pouch at 2–8 °C; do not freeze.

7. *Blocking buffer.* Bovine serum albumin, Cohn Fraction V has been found to be a suitable blocking protein for this application due to its high affinity for polystyrene surfaces. Additional blocking buffers to consider such as casein protein in carbonate–bicarbonate buffer, pH 9, or nonprotein blocking buffers and stabilizing agents are available from Surmodics (www.surmodics.com) or NOF's Biolipidure (www.nof.co.jp).

8. *Enzyme conjugate.* This depends upon which antibody type is to be monitored for the allergenic response, for example, specific IgE (sIgE) for allergy or specific IgG (total IgG or sIgG4) for food-intolerance studies. In any case, the corresponding HRP conjugate is to be used. This would be an antihuman antibody, preferably F_c-specific, for example, HRP-conjugated mouse antihuman IgG.

9. *Enzyme substrate.* The colorimetric ELISA requires development of signals on individual spots that are quantified by image analysis. Therefore, the TMB substrate must be a blot grade precipitating substrate. Vendors offering suitable blotting formulations include Thermo Scientific (1-Step Ultra TMB-Blotting Solution, Cat# 37574) and Moss, Inc. (TMBM or TMBH).

10. *Plasma samples.* Use plasma obtained from EDTA-treated (lavender top) or citrate-treated (blue top) blood collection tubes. Positive control plasma reference samples specific for food

allergy are available from commercial sources such as online from PlasmaLabs International (https://plasmalab.com) or Access Biologicals (https://accessbiologicals.com). Volumes of 1–2 mL are suitable for most analysis needs. Many vendors can also supply information regarding the patient's reactivity toward various foods such as the immunoCAP IgE value (kU/L) for each allergen tested in the sample.

11. *Reader.* Since the assay is developed based upon detection of an array of colorimetric spots, it is necessary to use an imaging device. So, to be clear, the common ELISA plate reader that measures signal in bulk solution is not suitable. Devices that can be used for this purpose are digital microscopes, camera phones, and imaging-based scanners or microplate readers: colorimetric, CCD or CMOS imaging type (e.g., sciREADER CL2, Scienion AG).

12. *Food protein content.* It is helpful to know the expected level of protein (grams protein/100-g weight of food) prior to performing an extraction. If you know the scientific name or even the common name for the food, it is possible to find this information online or from the vendor.

13. *Homogenization.* Start with a minimal buffer volume at about 1 mL per gram weight of material to be blended. Add additional buffer as needed to produce a slurry that can be strained. Keep solutions on ice to reduce protease activity.

14. *Ammonium sulfate.* Calculator available at http://encorbio.com/protocols/AM-SO4.htm.

15. *Dialysis.* Dialyze against 1 L PBS buffer per 1–3 cassettes with three buffer changes: e.g., initially 2 h, followed by overnight (18–20 h) and final exchange for an additional 2 h, remove dialysate from cassettes and centrifuge (10,000 rpm, 10 min) to remove turbidity (clarify).

16. *Probes.* Food protein extract *dialysates* are further diluted in PBS buffer to adjust the concentration for printing to about 1 mg/mL. After an initial print assessment, adjust the print concentration downward to about 0.25–1 mg/mL as required to achieve strong signal without overload (causing large diameter spots, smearing, comet tailing, coalesced spots, or other unsuitable spot morphologies). *Controls* used in this protocol refer to human immunoglobulin purified standards either obtained from natural sources or by recombinant means. For colorimetric analysis, the standard curve range needs to be adjusted separately for IgG, IgG4, and IgE. Purified IgG can be printed from about 10–100 μg/mL input concentration to cover the detectable range, while IgE standards dispensed from about 200–6000 IU/mL are recommended. Adjustment for differences in vendor preparation, purity, aggregation, buffers, additives, and source is to be expected.

17. *Lidding*. The source plate and assay plates should be lidded when transporting to avoid contamination with lint or other debris from the lab. Plastics may carry considerable electrostatic charge and easily pick up debris. If the source plate has been previously sealed and stored under refrigeration, it is advisable first to allow the plate to reach room temperature. Carefully remove the seal and replace with a clean lid. Centrifugation of a sealed, refrigerated plate may cause a vacuum to develop in the wells. This can lead to suction of the well contents out of the plate as the seal is removed, leading to potential cross-contamination.
18. *Dispensing*. Involves aspiration of the probe (the protein extract) into a PCD. The aspiration step may be problematic for certain extracts due to differences in viscosity (from lipid material; mucus-like aggregates, etc.) or the presence of fine particulates. In this case, further dilution and/or microfiltration may be necessary to achieve a stable and printable solution. Additional in-line cleaning steps might also be required after dispensing of these crude extracts to avoid cross-contamination. Also of importance is dispensing under controlled RH conditions. RH is usually set between ~50% and 70% at ~25 ± 3 °C. This is to prevent clogging of the dispenser capillary due to solvent evaporation leaving buffer residue and adsorbed proteins. Humidity also helps lower the static electricity buildup with the system especially with plastic substrates in use. Avoid higher RH that can lead to bulk wetting out of the printed surfaces, thereby coalescing and/or enlarging of the print droplets.
19. *Curing*. The drying process is critical to the performance of the plate assay. The printed plate should be stored for at least 2 days in a sealed container with drying agent so that the extract droplets completely dry out and permit the residual protein spot (allergen) to strongly adhere to the well surface (curing). Once the curing process is complete, the plate can be blocked for use. Completed plates may be stored at room temperature and dry in an air-tight pouch for several days. For longer term storage, seal dry in pouch under refrigeration but do not freeze.
20. *Plasma dilutions*. Antibody concentrations in plasma vary depending upon the allergenic response and half-life. In this ELISA, the detection of IgE and IgG requires different levels of dilution. For IgE, dilution at 1:2 (v/v) to 1:4 (v/v) in diluent or blocking buffer is recommended, while total IgG, which is present in higher concentration, should be diluted at least 1:100 (v/v); plasma IgG4 levels are lower; therefore, dilution at 1:20 (v/v) is recommended. It may be necessary to adjust the dilutions to improve sensitivity and reduce non-specific binding for different plasma samples.

21. *Static incubation.* This assay is run under static conditions to avoid the undesirable dispersion of the enzyme substrate precipitate (smearing) from the arrayed spots.

22. *Conjugate concentration.* The enzyme conjugate level is based upon the binding affinity of the antibody and the enzyme activity. Adjust concentration as recommended by the vendor or undertake a titration to determine the optimal dilution. In this case, 1:1000 (v/v) dilution represented about 1 μg/mL of enzyme conjugate.

23. *Signal development.* The enzyme-catalyzed generation of the TMB-colored precipitate is usually visible within a few minutes following the addition of substrate. The accumulation of precipitate may persist for 1 h or longer until the substrate is fully consumed. Longer development intervals increase background and may lead to dispersion of excess precipitate from the spot (sluffing effect).

24. *Reading.* Depending upon the optical geometry of the reader, the plate may be focused and read from the top or bottom of the well. For bottom reading, it is unnecessary to remove the TMB substrate solution from the well. For top reading, there may be problems with light diffraction through the liquid interface; therefore, it is recommended to remove the bulk solution from the well prior to reading. To reduce electrostatic, buildup, and attraction of lint onto the plate, wipe the well bottom with a little alcohol (e.g., 70% isopropyl alcohol).

25. Figure 2. The human IgG controls (purified) served two roles. First, location of the higher concentrated IgG in the corners (registration markers) of the array was utilized for spot recognition by the reader software's grid overlay. This enabled automated calculation of the pixel intensities for all spots of the array. Printing of a dilution series of control IgG further provided a dose–response curve from which to estimate the plasma IgG loading on each spot by interpolation.

    In this example, background signal can be observed on most spots of the array. This is most apparent with analysis of total IgG from plasma while IgG4 and IgE backgrounds are greatly reduced. Additional wash steps may be required to reduce background. A further dilution of the sample, the enzyme, or reduction in assay time might also be considered.

26. Figure 3. The standard curve is generated from the pixel intensities associated with spots of serially diluted total human IgG. The amount (mass) immobilized in spots has not been determined so the measure is indirect, and the interpolations of the food-specific IgG are semi-quantitative. However, dispense droplet volumes of the serially diluted IgG controls are measured during dispense such that the level of adsorbed

immunoglobulin in each spot can be estimated. Based upon 1.5 nL droplets dispensed to the surface at input IgG concentrations of 3.9–125 ng/μL, about 6–188 fg/spot can be calculated.

References

1. https://www.fda.gov/food/food-labeling-nutrition/food-allergies
2. https://www.fda.gov/food/buy-store-serve-safe-food/food-allergies-what-you-need-know
3. David NA, Penumarti A, Burks AW et al (2017) Food allergen extracts to diagnose food-induced allergic diseases: how they are made. Ann Allergy Asthma Immunol 119:101–107
4. Wiltshire S, O'Malley S, Lambert J, Kukanskis K, Edgar D et al (2000) Detection of multiple allergen-specific IgEs on microarrays by immunoassay with rolling circle amplification. Clin Chem 46(12):1990–1993
5. Lebrun SJ, Petchpuda WN, Huia A, McLaughlin CS (2005) Development of a sensitive, colorometric microarray assay for allergen responsive human IgE. J Immunol Methods 300:24–31
6. Vigh-Conrad KA, Conrad DF, Preuss D (2010) A protein allergen microarray detects specific IgE to pollen surface, cytoplasmic, and commercial allergen extracts. PLoS One 5(4):e10174. https://doi.org/10.1371/journal.pone.0010174
7. Feyzkhanova G, Voloshin S, Smoldovskaya O, Arefieva A et al (2017) Development of a microarray-based method for allergen-specific IgE and IgG4 detection. Clin Proteomics 14(1):1–13. https://doi.org/10.1186/s12014-016-9136-7
8. Matson RS (2016) Demonstration of a multiplex milk allergen ELISA using oligonucleotide-tethered principal component proteins. SOJ Biochem 2(1):1–7

Chapter 14

Mycotoxin Quantification by Competitive ELISA

Bryn E. Merrill and Robert S. Matson

Abstract

Enzyme-linked immunosorbent assays (ELISAs) are useful in the quantification of antigens like mycotoxins. The zearalenone (ZEA) mycotoxin is commonly found in cereal crops such as corn and wheat, which are used in domestic and farm animal feed. When consumed by farm animals, ZEA can cause harmful reproductive effects. The procedure used to prepare corn and wheat samples for quantification is described in this chapter. An automated procedure was developed to prepare samples from known levels of ZEA in corn and wheat. The final corn and wheat samples were analyzed using a competitive ELISA specific to ZEA.

Key words Competitive ELISA, Antigen, Antibody, Zearalenone, Mycotoxin, Colorimetric spectrophotometry

1 Introduction

In this chapter, we explore one of the many applications of enzyme-linked immunosorbent assays (ELISAs) that have been developed since their creation in 1971 [1]. The application of this study utilizes the competitive ELISA format. Competitive binding occurs between antigen in sample and a reagent antigen labeled with enzyme for binding to a capture antibody adsorbed to the surface of a microtiter plate [2] as described in Fig. 1. Free-floating antigens are washed away before the enzyme substrate used to assess their concentration is added, meaning that the higher the concentration of antigen in the sample, the fewer enzyme-labeled antigen competitors may be bound to the adsorbed antibody. The principles of competitive ELISA allowed us to quantify enzymatic transformations in samples of unknown antigen concentration.

1.1 Mycotoxin Quantification with Competitive ELISA

Our species of interest was zearalenone (ZEA), also known as F-2 mycotoxin or RAL. ZEA is a harmful estrogenic mycotoxin produced by certain fungal genera commonly found in corn, wheat, and other cereal crops [3]. This mycotoxin causes negative health

Robert S. Matson (ed.), *ELISA: Methods and Protocols*, Methods in Molecular Biology, vol. 2612,
https://doi.org/10.1007/978-1-0716-2903-1_14,

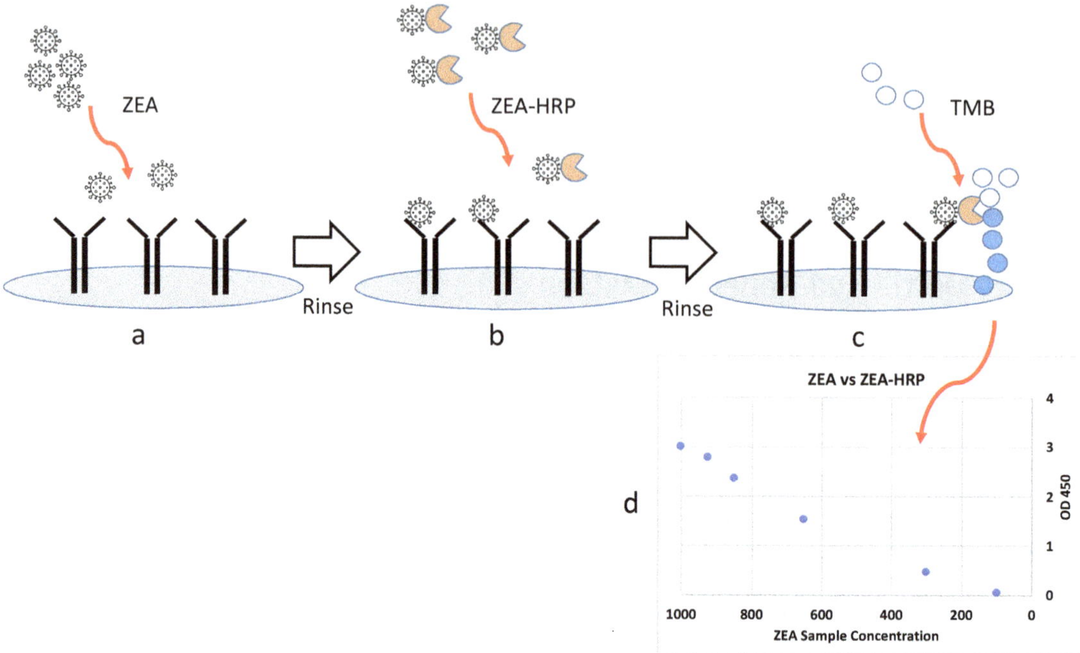

Fig. 1 Zearalenone (ZEA) competitive ELISA. ZEA extracted from grain samples is added to well coated with an anti-ZEA capture antibody (**a**). After a rinse, ZEA conjugated with HRP enzyme is added (**b**). The conjugate will be bound by any unoccupied capture antibody. Following an additional rinse, TMB substrate is added, resulting in a blue color (yellow upon acidification) indicating the presence of ZEA-HRP (**c**). The color intensity is proportionate to the level of bound ZEA-HRP or inversely proportional to the ZEA in the sample. So, high levels of ZEA in the sample will result in a decrease in color absorbance

effects in domestic and farm animals, including reproduction issues and infertility. The goal of this experiment is to provide a quick and efficient means of detecting harmful levels of ZEA in cereal crops [3]. In this study, we first evaluated and then optimized Apricot Designs' instrumentations to be used in work on the solid-phase extraction (SPE) processes that would transfer the mycotoxins in the samples from the solid cereal crops to a liquid phase. The amount of extracted ZEA was then estimated via a competitive ELISA from HELICA Biosystems, which is coated with an antibody optimized to bind with free ZEA from the sample [3]. The HRP-ZEA conjugate binds with any remaining unbound antigen in the well before the substrate is added. The TMB substrate turns blue in the presence of enzyme; therefore, the more enzymes that are bound to the well, then the more intense the blue color. Conversely, the more ZEA in the sample, the less intense the color as these unbound antigens would be washed away before the substrate was added. Using this correlation and a standard curve, the levels of ZEA in corn and wheat samples are determined through colorimetric spectrophotometry.

2 Materials

2.1 Extraction

1. Solid samples: Ground corn and wheat (*see* **Note 1**).
2. Solvent: 70% methanol.
3. 2.0-mL centrifuge tubes.
4. Waters Oasis Prime HLB Sorbent column (*see* **Note 2**).
5. EVEN96 Extractor (Apricot Designs).

2.2 Evaporation

1. 96-Well deep well plate.
2. Compressed N_2 gas (*see* **Note 3**).
3. Solvent: 70% methanol.
4. ENVX96 or EVX192 Evaporator (Apricot Designs).

2.3 Colorimetric ELISA Development

1. HELICA Zearalenone ELISA Assay—Low Matrix kit. Cat# 981ZEA01LM-96 (*see* **Note 4**).
2. Automated Multi-Pipettor, iPipettePro (Apricot Designs).

3 Methods

A total of 12 samples were prepared using three extraction techniques (*see* **Note 5**) and analyzed following the flowchart in Fig. 2.

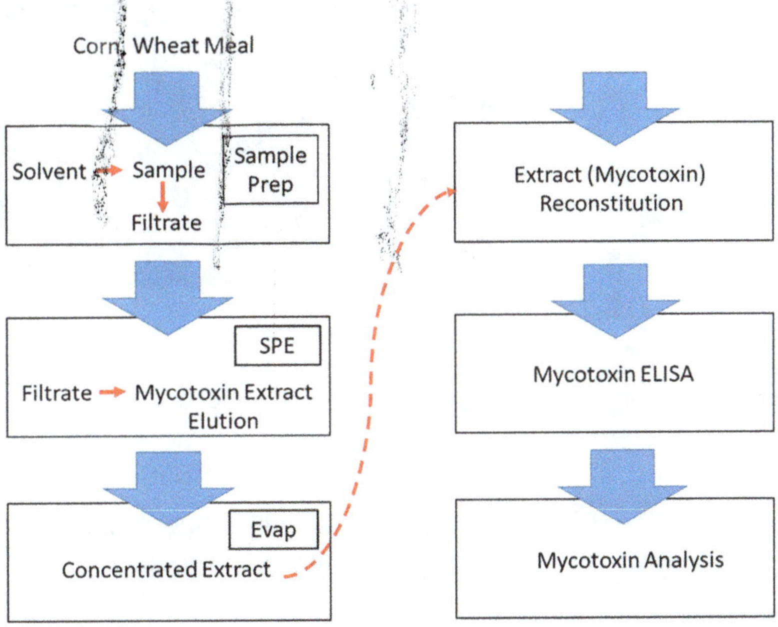

Fig. 2 Semi-automated sample process flowchart

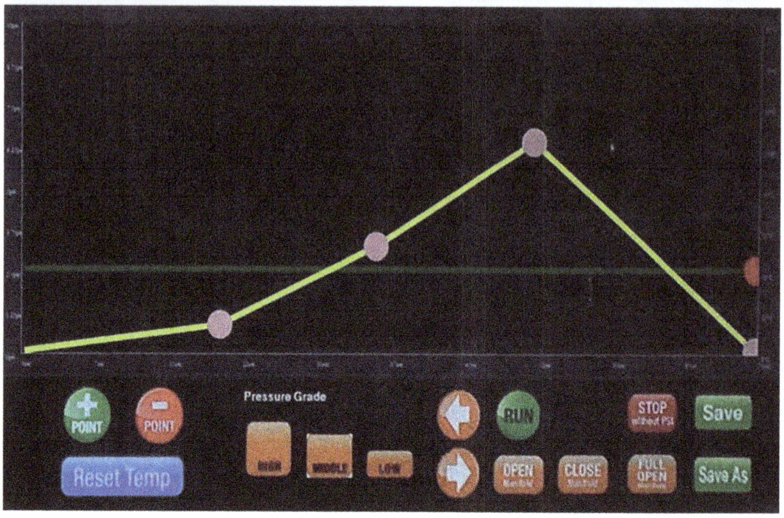

Fig. 3 Pressure profile for the conditioning of the column with methanol

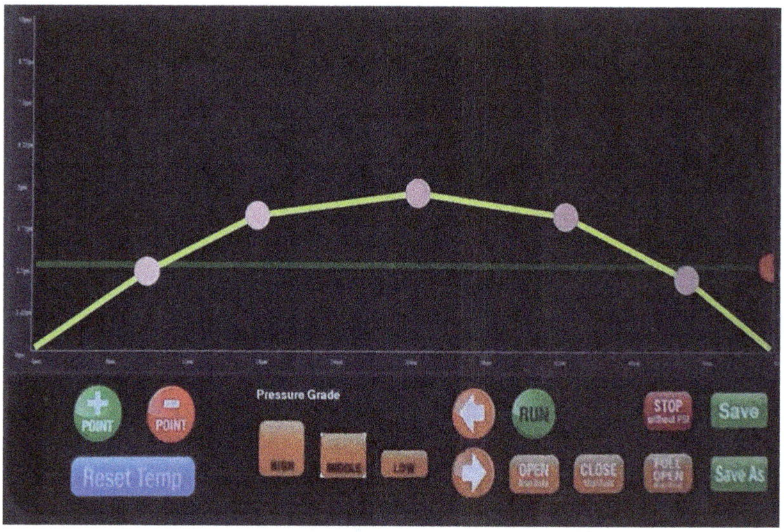

Fig. 4 Pressure profile for the equilibration of the column with DI water

3.1 Extraction

1. Prepare ZEA samples at about 0.2 g/mL in methanol.
2. Set samples on a mixer for 30 min, and then let settle for 10 min at room temperature (*see* **Note 5**).
3. Collect the liquid sample via pipette for extraction (*see* **Note 6**).
4. The column was first conditioned with 500 μL of 100% methanol using the pressure profile depicted as in Fig. 3.
5. Next, 500 μL of deionized (DI) water was added to the column under equilibration pressure profile (Fig. 4).

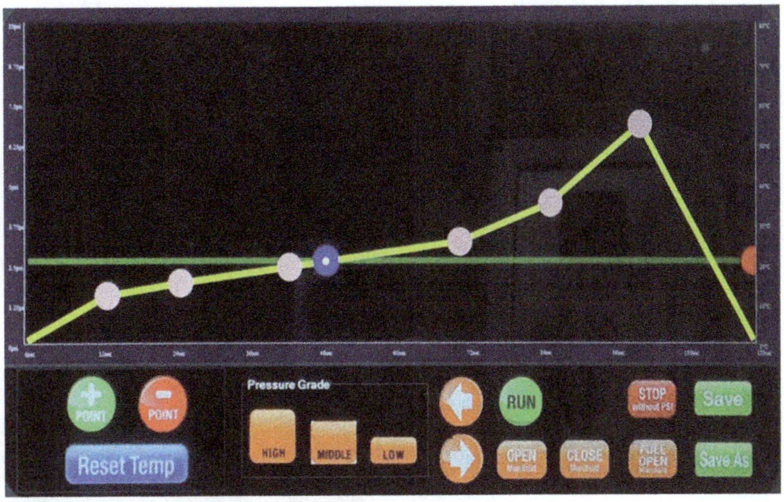

Fig. 5 Pressure profile for loading the sample

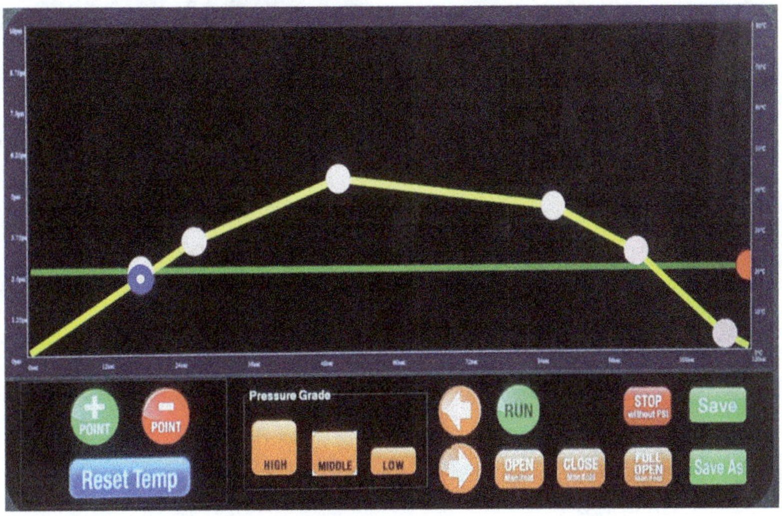

Fig. 6 Pressure profile for the elution of the sample with methanol

6. The sample was then loaded to the column, 50 μL to each well, and pressurized with the profile shown in Fig. 5.
7. The final elution step is performed using 500 μL of 100% methanol with the elution pressure profile shown in Fig. 6.

3.2 Evaporation

The eluate collected from the EVEN96 is then evaporated with the Apricot Designs' EVX96 (Fig. 7).

1. Nitrogen gas (N_2) is heated in the EVX96 to increase the rate of evaporation as shown by the data in Fig. 8 (*see* **Note 6**).

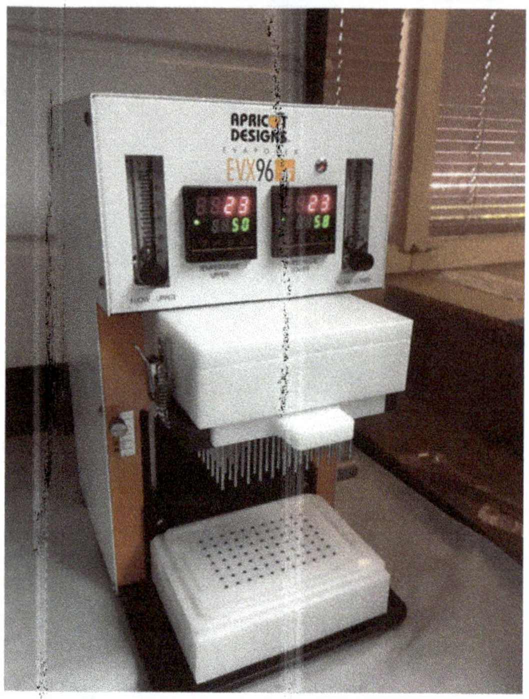

Fig. 7 The EVX96 used for the second experiment using methanol solutions to optimize evaporation techniques

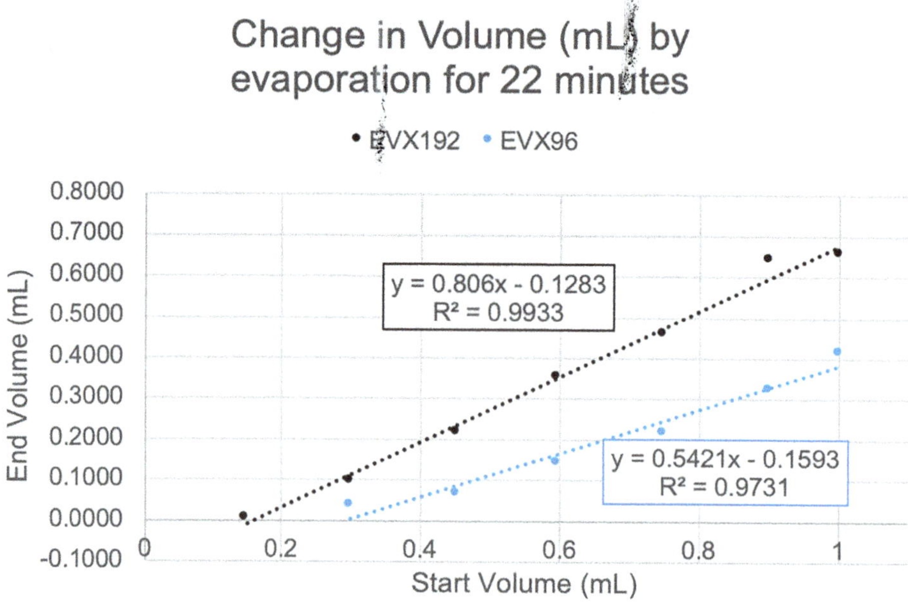

Fig. 8 Comparison of evaporation rates for methanol between the EVX192 (black) and the EVX96 (blue)

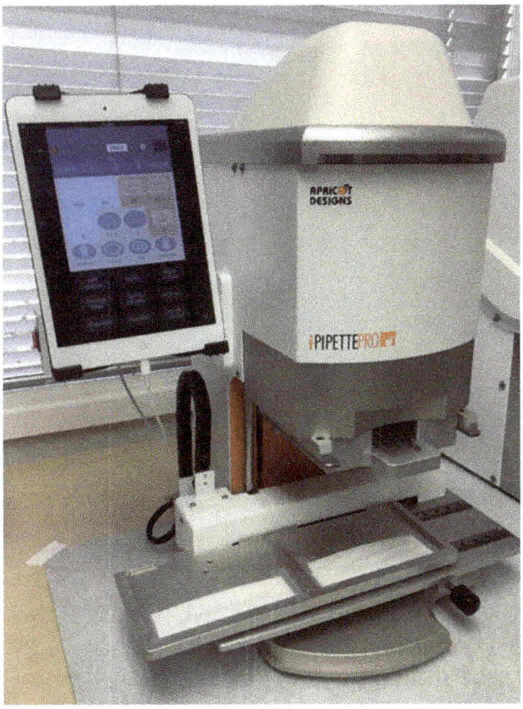

Fig. 9 The iPipettePro used in our experimentation

2. The evaporated samples were then re-solvated with 100 μL of 70% methanol on the Apricot Designs' iPipette (Fig. 9).
3. The resulting solutions were then diluted in 100 μL of sample (or calibrator) to 200 μL of diluent using the iPipette (Fig. 10).

3.3 Colorimetric ELISA Development

The samples were mixed in the HELICA Zearalenone Assay preparation plate using the iPipette (Figs. 9 and 10) before being transferred to the ELISA plate for the assay.

1. 100 μL of the solution from the mixing plate was transferred to the ELISA plate, which was then incubated for 10 min on a mixer.
2. The wells are next rinsed three times with PBST followed by the addition of 100 μL of HRP enzyme to each well.
3. The plate is incubated another 10 min before the wells are emptied and rinsed again.
4. 100 μL of TMB substrate is added and a third 10-min incubation is performed.
5. 100 μL of a stop solution of sulfuric acid was added to the ELISA plate.
6. The plate was then analyzed at 450 nm in a DYNEX spectrophotometer. Calibrator data were used to create the standard

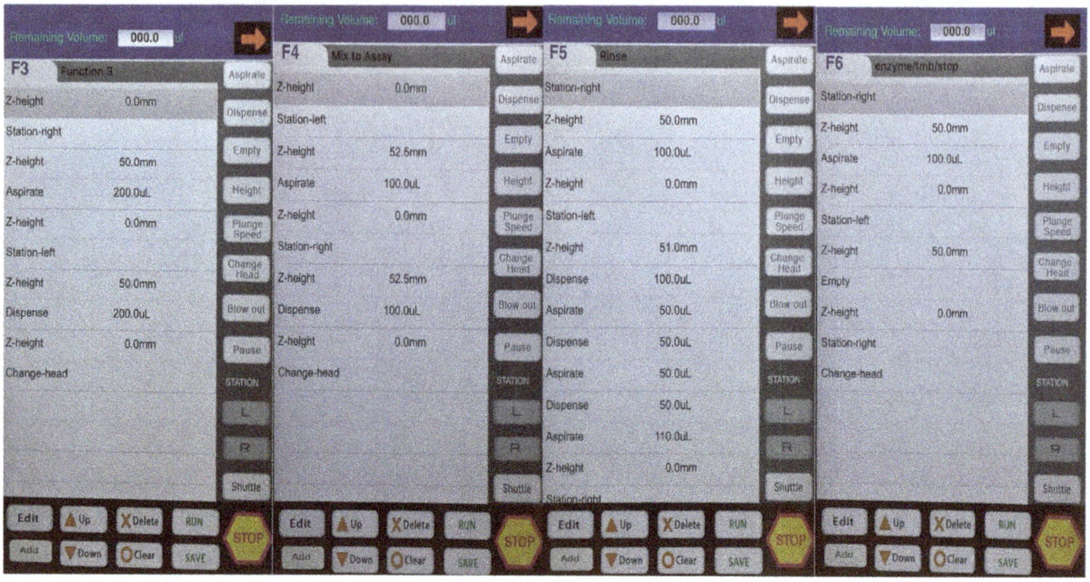

Fig. 10 iPipette functions 3–6 for running ELISA. Comprising transferring of the diluent to the HELICA mixing plate, introducing the samples and standards, transferring the solution to the ELISA plate, rinsing, and adding enzyme, substrate and stop solution.

curve in Fig. 11, which was compared to the sample data recorded in Table 1. The sample data collected were used to compare the extraction processes in Figs. 12 and 13.

4 Notes

1. Preparation for extraction began with four finely ground corn and wheat samples from Trilogy (Washington, MO) with known ZEA concentrations; two nondetectable concentrations (one wheat, one corn), and two samples of known concentration (362.9 ppb ZEA corn sample, 237.8 ppb ZEA wheat sample).
2. The Waters Oasis Prime HLB column was coupled with the EVEN96 (Figs. 3, 4, 5, 6, 14, and 15). The prepared samples were extracted on Apricot Designs' EVEN96 (Fig. 14) using Waters Oasis Prime HLB Sorbent column according to the method shown in Fig. 15.
3. The evaporation process used large amounts of N_2 gas and took much more time for solutions with volumes greater than 50 μL; the evaporation process was deemed unnecessary for anything above 50 μL.
4. The HELICA Zearalenone Assay kit includes all the needed reagents for the assay including 96-well preparation plate,

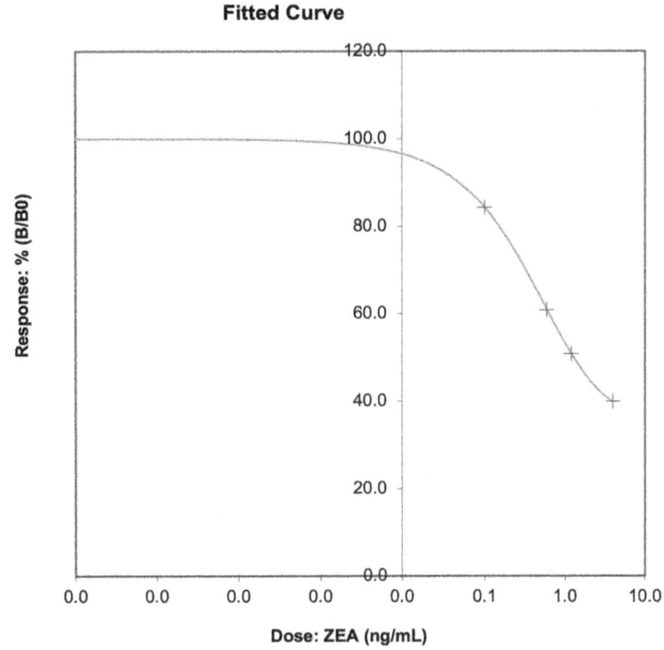

Fig. 11 Standard curve created using AssayFit for Excel and five of the standard ZEA solutions included in the ELISA Kit: 0, 0.1, 0.6, 1.2, and 4 ng/mL

Table 1
Extraction data and ZEA recovery after ELISA through different extraction processes labeled "single," "re-extract," "combined," and "5 days, 4°C"

Certified content		ZEA	Test	Extraction	Calculated	ZEA	ELISA
Grain sample	Sample ID	Sample PPB (ng/g)	Sample Wt g	process	Dose ng	Recovered PPB (ng/g)	Recovery %
Corn	CZ-1	362.9	0.203	Single	1.79	97	26.7
Corn	CZ-2	362.9	0.203	Single	1.18	65.7	18.1
Corn	CZ-2b	362.9	0.203	Re-extract	0.63	36.1	15
	CZ-2 total			Combined	1.81	101.8	28
Corn	CZ-3	362.9	0.204	5 days, 4°C	1.79	123.1	51.8
Wheat	WZ-1	237.8	0.201	Single	1.18	66.3	27.9
Wheat	WZ-2	237.8	0.205	Single	0.35	19.2	8.1
Wheat	WZ-2b	237.8	0.205	Re-extract	0.56	31.8	13.4
	WZ-2 total			Combined	0.91	51	21
Wheat	WZ-3	237.8	0.204	5 days, 4°C	1.98	156.5	65.8
Corn	CND-1	<50	0.215	Single	0.03	1.5	ND
Wheat	WND-1	<50	0.203	Single	0.05	2.9	ND

Certified "zero" control was reported at below the limit of detection at <50 PPB. *ND* not determined

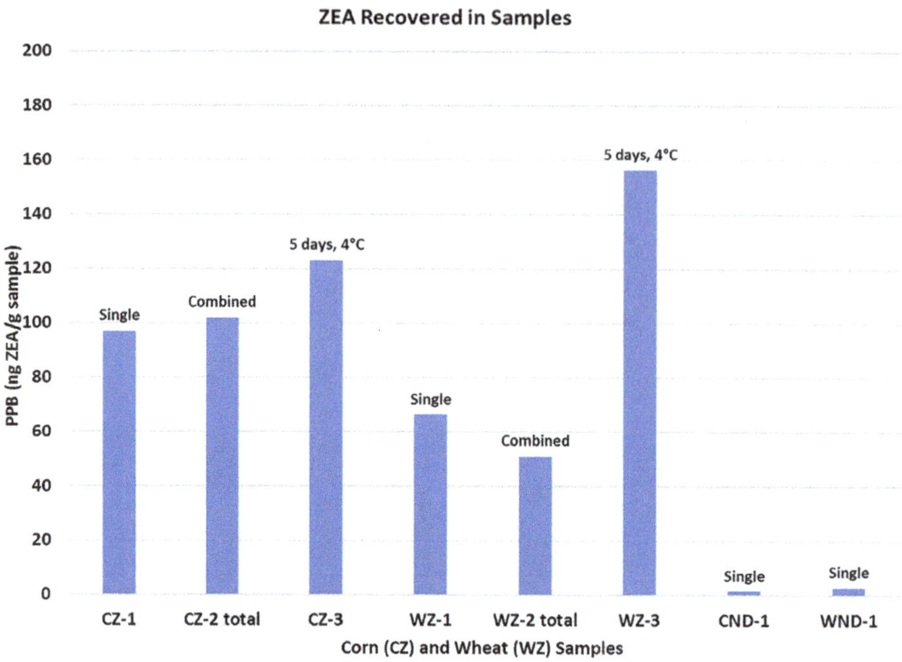

Fig. 12 The mass of ZEA recovered in ppb for each tested sample corresponding to the different methods

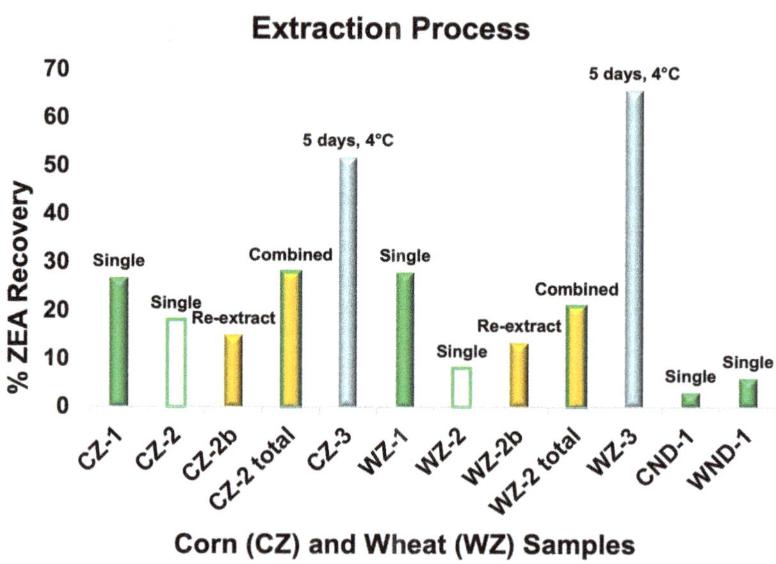

Fig. 13 Percent recovery of each sample highlighting the differences in detected ZEA with different methods

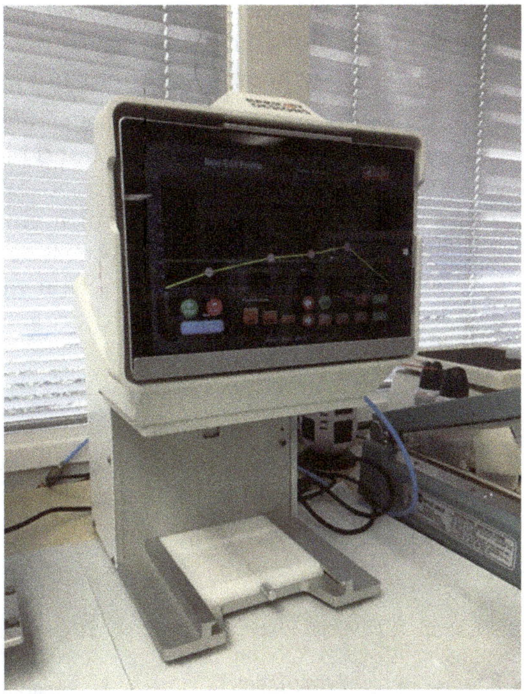

Fig. 14 The EVEN96 instrument from Apricot Designs used in our experimentation

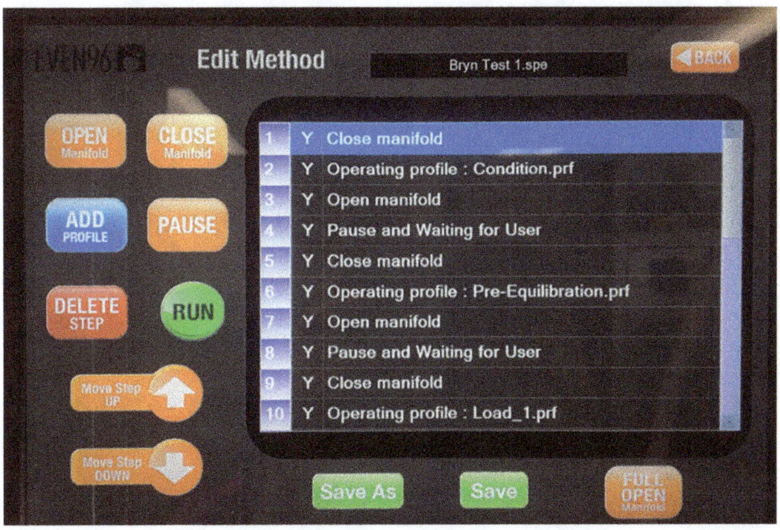

Fig. 15 With each addition of solvent, sample, or solution, the column is pressurized with designated pressure profiles

96-well antigen-coated plate, Zearalenone standards, Zearalenone HRP-conjugate, assay diluent, substrate reagent (TMB), acidic stop solution, and PBST washing buffer.

5. In a process we call "single," the liquid solution is then pulled off the solid sample and placed into a different 2.0-mL centrifuge tube. The process called "re-extract" uses a sample that has already been through the "single" process. 1 mL of 70% methanol–water was added to the same solid sample used in "single" and was set on the mixer for 30 min with 10 min to settle. In the third process, "5 day, 4°C," a new sample was prepared in the same way as the "single" process, but the settle time was extended to 5 days at 4 °C.
6. After the samples were allowed to settle, the liquid layer was carefully pulled from the top of the samples, leaving the solid behind in the centrifuge tube.

References

1. Engvall E, Perlmann P (1972) Enzyme-linked immunosorbent assay, Elisa: III. Quantitation of specific antibodies by enzyme-labeled anti-immunoglobulin in antigen-coated tubes. J Immunol 109(1):129–135
2. Crowther JR (1995) ELISA: theory and practice. Springer, Berlin
3. Zearalenone ELISA assay-low matrix. Cat. No. 981ZEA01LM-96, Ver. 1. HELICA Biosystems, Inc., Santa Ana. May 2018

Chapter 15

Lumit: A Homogeneous Bioluminescent Immunoassay for Detecting Diverse Analytes and Intracellular Protein Targets

Byounghoon (Brian) Hwang, Juliano Alves, Dan Lazar, Nidhi Nath, Laurie Engel, Martha O'Brien, Kevin Hsiao, Kevin Kupcho, Becky Godat, Rod Flemming, Said Goueli, and Hicham Zegzouti

Abstract

Traditional immunoassays to detect secreted or intracellular proteins can be tedious, require multiple washing steps, and are not easily adaptable to a high-throughput screening (HTS) format. To overcome these limitations, we developed Lumit, a novel immunoassay approach that combines bioluminescent enzyme subunit complementation technology and immunodetection. This bioluminescent immunoassay does not require washes or liquid transfers and takes less than 2 h to complete in a homogeneous "Add and Read" format. In this chapter, we describe step-by-step protocols to create Lumit immunoassays for the detection of (1) secreted cytokines from cells, (2) phosphorylation levels of a specific signaling pathway node protein, and (3) a biochemical protein–protein interaction between a viral surface protein and its human receptor.

Key words Immunoassay, Bioluminescence, Cytokine, Phosphorylation, Antibody, NanoBiT, Homogeneous, SARS-CoV2

1 Introduction

There are multiple methods for performing immunoassays, including direct, indirect, or sandwich formats. Moreover, immunoassays can be separative (heterogeneous) with multiple washing steps, or nonseparative (homogeneous), where the detection is performed in solution without washing or antibody/analyte capture on the plate [1]. The Lumit Immunoassay platform is a novel bioluminescent no-wash and homogeneous method based on NanoLuc Binary Technology (NanoBiT) [2]. NanoLuc is an engineered 19 kDa

Robert S. Matson (ed.), *ELISA: Methods and Protocols*, Methods in Molecular Biology, vol. 2612,
https://doi.org/10.1007/978-1-0716-2903-1_15,

luciferase enzyme derived from the deep-sea luminous shrimp *Oplophorus gracilirostris* that is relatively stable and produces very bright and sustained bioluminescence [3]. The NanoBiT is a structural complementation reporter ideal for protein–protein interaction (PPI) studies [4–6]. The NanoBiT system is based on NanoLuc enzyme, and it is composed of two subunits, Large BiT (LgBiT; 18 kDa) and Small BiT (SmBiT; 11 amino acid peptide), that can be expressed as fusions with or chemically conjugated to target proteins of interest. The LgBiT and SmBiT subunits are stable and have minimal self-association because the SmBiT peptide and the LgBiT protein associate with relatively low affinity (Kd = 190 μM) [2]. Therefore, only when two target proteins tagged with these subunits interact are the subunits close enough to form an active enzyme that generates light in the presence of its substrate furimazine. In one Lumit format, the NanoBiT subunits are conjugated to two primary antibodies that recognize two different epitopes on a target protein. When the protein is present in the sample to be analyzed, the two NanoBiT-conjugated antibodies bind to their corresponding epitopes on the target and the subunits are brought into proximity creating an assisted enzyme complementation that generates luminescence (Fig. 1a). The luminescent signal generated can be detected using a luminometer. In another Lumit format, NanoBiT-labeled secondary antibodies that recognize target primary antibodies bypass NanoBiT conjugation to the primary antibody [7]. In this indirect Lumit application, NanoBiT subunits are conjugated to a pair of secondary antibodies against two different species (Anti-Rabbit, Anti-Mouse, or Anti-Goat). The target protein is detected by adding an antibody mix containing two unlabeled primary antibodies against the target protein along with SmBiT- and LgBiT-conjugated secondary antibodies (Lumit secondary antibodies). Binding of the primary and Lumit secondary antibody complexes to their corresponding epitopes brings NanoBiT subunits into proximity to generate light in proportion to the amount of the target protein (Fig. 1b).

Here, we describe the method we developed to conjugate NanoBiT subunits to antibodies to create any Lumit Immunoassay. We then outline its utility in three different applications that use either labeled primary antibodies (direct Lumit) or labeled secondary antibodies (indirect Lumit). We also present detailed protocols for determining IL-4 cytokine released from cells in culture, detection of phosphorylated ERK1 in a cell lysate after activation of the MAPK signaling pathway, and detection of the SARS-CoV2 spike RBD and human ACE2 receptor PPI (Fig. 1).

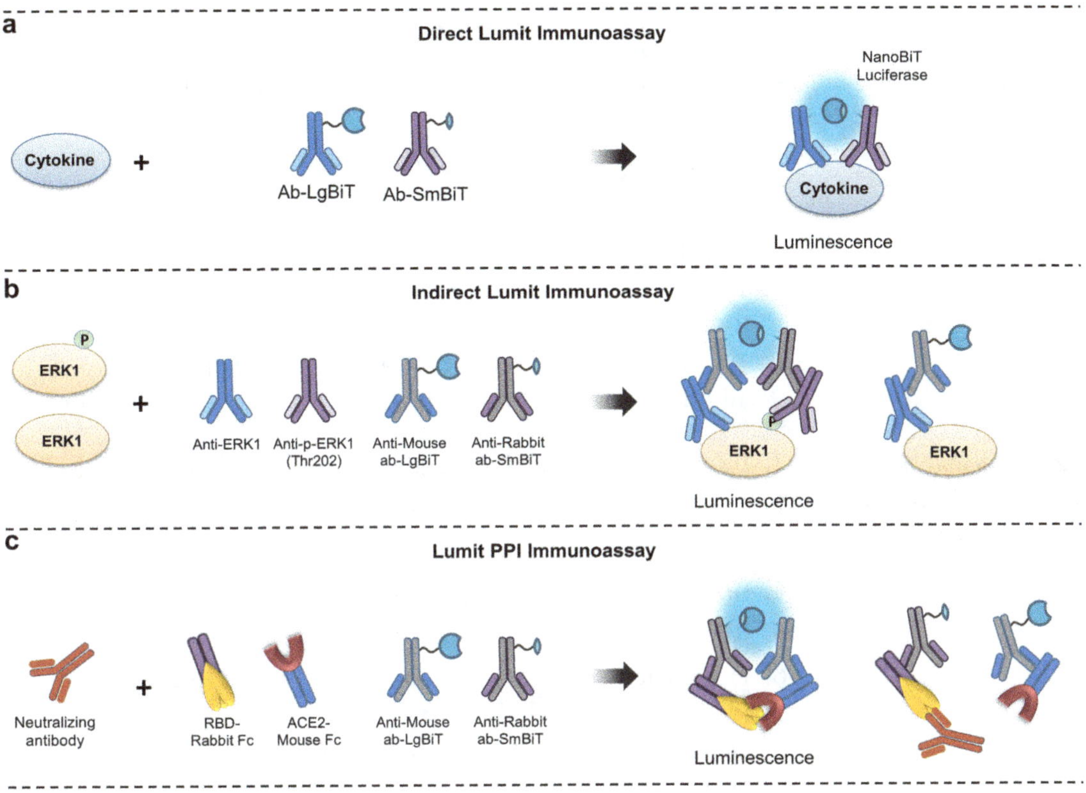

Fig. 1 Lumit Immunoassays principle. (**a**) Direct Lumit Immunoassay for the detection of cytokines. This method uses two NanoBiT-labeled primary antibodies to detect the target protein. (**b**) Indirect Lumit Immunoassay for the detection of phosphorylated target in cell lysate. Phosphorylated target protein in lysed cells is recognized by unlabeled primary antibody pair. The NanoBiT conjugated secondary antibodies then recognize their cognate primary antibodies, forming the NanoBiT enzyme and generating luminescence. (**c**) Lumit PPI immunoassay. This method uses NanoBiT-labeled antibodies against the tags of the two recombinant proteins that are involved in the protein–protein interaction. Here, the NanoBiT conjugated secondary antibodies (Anti-Mouse and Anti-Rabbit) are used to bind the rabbit and mouse Fc tags of the interacting SARS-CoV2 RBD and human ACE2 proteins, respectively

2 Materials

2.1 Labeling of Antibodies with NanoBiT Subunits

1. 1 mg/mL antibody solution (*see* **Note 1**).
2. Lumit™ Immunoassay labeling kit (Promega). One kit includes.
   - 20 mM HaloTag® Ligand.
   - IGEPAL® CA-630 (100%).
   - 20 μM Lumit™ HaloTag®-LgBiT.
   - 20 μM Lumit™ HaloTag®-SmBiT.
3. Zeba™ Spin Desalting Columns 7 K MWCO (Thermo Scientific), 0.5 mL or equivalent.

4. 10 mM sodium bicarbonate buffer (pH 8.3–8.5) (user supplied).
5. Phosphate-buffered saline (PBS) (user supplied).
6. Glycerol (Sigma).
7. Magne® HaloTag® Beads (Promega).
8. MagneSphere® Technology Magnetic Separation Stands, 1.5 mL (Promega).
9. 10% IGEPAL® CA-630 solution in PBS. Make fresh prior to use.
10. PBS containing 0.05% IGEPAL® CA-630. Prepare 5 mL by adding 25 μL of 10% IGEPAL® CA-630 to 5 mL of PBS.

2.2 Detection of Cytokine Release from Cells in Culture

1. Purified monoclonal antibodies to human IL-4 (*see* Table 1).
2. Recombinant human IL-4 (PeproTech).
3. 1× phosphate-buffered saline (PBS).
4. Bovine serum albumin (BSA) (Promega).
5. RPMI 1640 cell culture medium (Gibco).
6. Heat-inactivated fetal bovine serum (FBS) (Gibco).
7. Lumit™ Immunoassay Detection Reagent B (Promega), including Lumit™ Detection Substrate B and Lumit™ Detection Buffer B.

Table 1
Antibodies screened for the development of Lumit hIL-4 Immunoassay

Assigned screen I.D.	Target	Clone ID	Vendor	Catalog #	Host species
mAb1	Human IL-4	25D2	Thermo Fisher Scientific	M450	Rat
mAb2	Human IL-4	860A4B3	Thermo Fisher Scientific	AHC0642	Mouse
mAb3	Human IL-4	MP4-25D2	Thermo Fisher Scientific	50-136-07	Rat
mAb4	Human IL-4	E10021	Thermo Fisher Scientific	MIL4I	Mouse
mAb5	Human IL-4	E10022	Thermo Fisher Scientific	M2IL4I	Mouse
mAb6	Human IL-4	E10023	Thermo Fisher Scientific	M3IL4I	Mouse

8. Human T helper 2 (Th2) cells (*see* **Note 2**).
9. Human peripheral blood mononuclear cells (PBMCs) (BioIVT).
10. Cell Stimulation Cocktail, 500× (eBioscience).

2.3 Detection of Phosphorylated ERK1 (p-Thr202) in Cell Lysate

1. MCF-7 cells (ATCC).
2. Dulbecco's Modified Eagle Medium (DMEM) (Gibco Thermo Fisher Scientific).
3. Phenol-free DMEM (Gibco Thermo Fisher Scientific).
4. FBS (Avantor).
5. Penicillin/streptomycin (Gibco, 10 mg/mL).
6. Phorbol 12-myristate 13-acetate (PMA) powder (Promega).
7. Human epidermal growth factor (EGF) powder (Promega).
8. Human hepatocyte growth factor (HGF) powder (R&D Systems).
9. Insulin (Sigma).
10. U0126 powder (Promega).
11. Dimethylsulfoxide (DMSO).
12. Dulbecco's phosphate-buffered saline (DPBS) (Gibco).
13. Rabbit anti-p-ERK1 (Thr202) antibody (Abcam).
14. Mouse anti-ERK1 antibody (R&D Systems).
15. Lumit™ Immunoassay Cellular System—Set 1. One kit includes.
    - 50 μL Digitonin, 2%.
    - 1.5 mL Lumit™ Immunoassay Buffer C, 10 × .
    - 200 μL Lumit™ Substrate C.
    - 30 μL Lumit™ Anti-Mouse Ab-LgBiT.
    - 30 μL Lumit™ Anti-Rabbit Ab-SmBiT.
    - 1 mL antibody dilution buffer.
    - 10 μL fluorogenic live cell substrate (GF-AFC substrate), 100 mM.
16. Complete growth medium for MCF-7 cells: Add 50 mL of FBS and 5 mL of penicillin–streptomycin to 500 mL of DMEM and store at 4 °C until use.
17. Starvation medium: Add 5 mL of penicillin–streptomycin to 500 mL of phenol-free DMEM and store at 4 °C until use.
18. 5 mM PMA stock: Add 1.62 mL of DMSO to 5 mg of the PMA powder. Aliquots are stored at −20 °C.
19. 100 μg/mL EGF stock: Add 1 mL of water to 100 μg of the EGF powder. Aliquots are stored at −20 °C.

20. 100 μg/mL HGF stock: Add 50 μL of sterile DPBS containing 0.1% BSA to 5 μg of the HGF powder. Aliquots are stored at –20 °C.
21. 1.72 mM insulin is stored at 4 °C.
22. 10 mM U0126 stock: Add 234 μL of DMSO to 1 mg of the U0126 powder and shake vigorously for 30 seconds. Aliquots are stored at −20 °C.
23. Halt™ Protease and Phosphatase Inhibitor Cocktail, EDTA-free (100×) (Thermo Fisher Scientific).
24. EGF stimulation medium 1 (100 ng/mL EGF): Add 2 μL of EGF stock to 2 mL of starvation medium.
25. EGF stimulation medium 2 (800 ng/mL EGF): Add 2 μL of EGF stock to 248 μL of starvation medium.
26. PMA stimulation medium (1 μM PMA): Add 1 μL of PMA stock to 5 mL of starvation medium.
27. HGF stimulation medium (100 ng/mL HGF): Add 2 μL of HGF stock to 2 mL of starvation medium.
28. Insulin stimulation medium (1 μM insulin): Add 1.16 μL of insulin stock to 2 mL of starvation medium.
29. GF-AFC medium (50 μM): Add 2.5 μL GF-AFC to 5 mL of starvation medium.
30. GF-AFC/U0126 (50/10 μM) medium: Add 2 μL U0126 stock to 2 mL of GF-AFC medium.
31. GF-AFC/DMSO medium: Add 2 μL DMSO to 2 mL of GF-AFC medium.
32. Anti-p-ERK1 antibody working stock: Make 100 μL of a 50 μg/mL solution by diluting appropriate amount (Lot specific) of the rabbit anti-p-ERK1 (Thr202) antibody in the antibody dilution buffer.
33. Anti-ERK1 antibody working stock: Make 100 μL of a 50 μg/mL solution by diluting appropriate amount (Lot specific) of the mouse anti-ERK1 antibody in the antibody dilution buffer.
34. Immunoassay reaction buffer: Mix 1 mL of Lumit Immunoassay Buffer C with 100 μL Halt protease and phosphatase inhibitors cocktail and 9 mL water.
35. Lysis solution (0.1% digitonin): Prepare 1.2 mL of lysis solution by adding 60 μL digitonin (2%) to 1.14 mL of immunoassay reaction buffer.
36. Lumit p-ERK antibody mix: Mix 24 μL of anti-p-ERK1 antibody working stock, 24 μL of anti-ERK1 antibody working stock, 24 μL of Lumit Anti-Mouse Ab-LgBiT, and 24 μL of Lumit Anti-Rabbit Ab-SmBiT with 3.9 mL of 1× immunoassay reaction buffer. This volume is enough for 80 assay wells (*see* **Notes 3–5**).

37. Lumit reagent control mix: Mix 9 μL of Lumit Anti-Mouse Ab-LgBiT and 9 μL of Lumit Anti-Rabbit Ab-SmBiT with 1.48 mL of 1× immunoassay reaction buffer. This volume is enough for 30 assay wells (*see* **Notes 3–5**).
38. Lumit Detection Reagent: Prepare 2.5 mL of fresh Lumit Detection Reagent by mixing 200 μL Lumit Substrate C with 2.3 mL immunoassay reaction buffer (1:12.5 dilution) just before use. This volume is enough for 100 assay wells (*see* **Notes 3–5**).

2.4 Detection of SARS-CoV2 Spike RBD: Human ACE2 Receptor Protein–Protein Interaction

1. 1 mg/mL (19.88 μM) SARS-CoV-2 RBD (rabbit Fc) (Sino Biological). Prepare by mixing 100 μg lyophilized protein with 100 μL dilution buffer (10 mM HEPES pH 7.5, 150 mM NaCl, 0.1% BSA, 50% glycerol).
2. 1 mg/mL (9.09 μM) human ACE2 (mouse Fc) (Sino Biological). Prepare by mixing 100 μg lyophilized protein with 100 μL dilution buffer (10 mM HEPES pH 7.5, 150 mM NaCl, 0.1% BSA, 50% glycerol).
3. Lumit™ Anti-Mouse Ab-LgBiT (Promega).
4. Lumit™ Anti-Rabbit Ab-SmBiT (Promega).
5. Lumit™ Detection Substrate C (Promega).
6. Lumit™ Immunoassay Buffer C, 10× (Promega).
7. 0.5 μM RBD-rFc stock solution. Prepare 40 μL by mixing 1 μL of 1 mg/mL SARS-CoV-2 RBD (rabbit Fc) with 39 μL dilution buffer (10 mM HEPES pH 7.5, 150 mM NaCl, 0.1% BSA, 50% glycerol) (*see* **Notes 6** and 7).
8. 0.5 μM ACE2-mFc stock solution. Prepare 40 μL by mixing 2.2 μL of 1 mg/mL human ACE2 (mouse Fc) with 37.8 μL dilution buffer (10 mM HEPES pH 7.5, 150 mM NaCl, 0.1% BSA, 50% glycerol) (*see* **Notes 6** and 7).
9. 1× immunoassay reaction buffer: Mix 350 μL of Lumit™ Immunoassay Buffer C, 10× with 3150 μL water (*see* **Note 8**).
10. 7.5 nM SARS-CoV-2 RBD (rabbit Fc) solution: Mix 7.5 μL of RBD-rFc stock solution with 492.5 μL of 1× immunoassay reaction buffer (*see* **Note 8**).
11. 7.5 nM human ACE2 (mouse Fc) solution: Mix 7.5 μL of ACE2-mFc stock solution with 492.5 μL of 1× immunoassay reaction buffer (*see* **Note 8**).
12. Lumit Antibody Mix: Add 7.5 μL of Lumit™ Anti-Mouse Ab-LgBiT and 7.5 μL of Lumit™ Anti-Rabbit Ab-SmBiT to 1235 μL of 1× immunoassay reaction buffer (*see* **Notes 5, 7,** and **8**).

13. Lumit Detection Reagent: Mix 50 μL of Lumit™ Substrate C with 575 μL of 1× immunoassay reaction buffer (*see* **Notes 5, 7,** and **8**).
14. 6.67 μM (1 mg/mL) SARS-CoV/SARS-CoV-2 Spike antibody (Sino Biological 40,150-D002).
15. 6.67 μM (1 mg/mL) Purified anti-SARS-CoV-2 S Protein S1 Recombinant Antibody (BioLegend 938,602).
16. Five to ten verified pre-pandemic negative COVID-19 samples (*see* **Notes 9** and **10**).
17. Patient-derived samples to be tested (serum or plasma) (*see* **Notes 9** and **10**).
18. 50% diluted patient-derived sample: Mix 5 μL of sample to 5 μL 1× immunoassay reaction buffer.

2.5 Equipment

1. 96-Well white flat-bottom untreated polystyrene microplate.
2. White, tissue culture-treated, 96-well microplates.
3. Single-channel pipettes (0.2–2 μL, 20–200 μL, 100–1000 μL) and tips.
4. 1.5-mL microcentrifuge tubes (polypropylene).
5. Standard benchtop centrifuge.
6. Conical centrifuge tubes (50 mL, polypropylene).
7. Multichannel pipettes (20–200 μL, 100–1200 μL) and tips.
8. Orbital plate shaker.
9. Standard, microplate-reading luminometer or multimode plate reader.
10. Cell culture laminar flow hood, class II.
11. Humidified cell culture incubator allowing for standard cell culture conditions (37 °C, 5% CO_2).
12. Incubator or plate warmer at 22–25 °C.
13. Centrifuge compatible with multiwell plates.
14. SDS-PAGE gel such as 4–20% Criterion TGX gel (Bio-Rad) and apparatus.

3 Methods

3.1 Labeling of Antibodies with SmBiT and LgBiT

Here, we describe a novel chemical approach for labeling antibodies and proteins with NanoBiT subunits using HaloTag technology [8]. This two-step process (Fig. 2a) first chemically labels the antibody with HaloTag ligand (MW: 509 Da) using well-established amine chemistry. In the second step, purified HaloTag-NanoBiT fusion protein is added to the activated

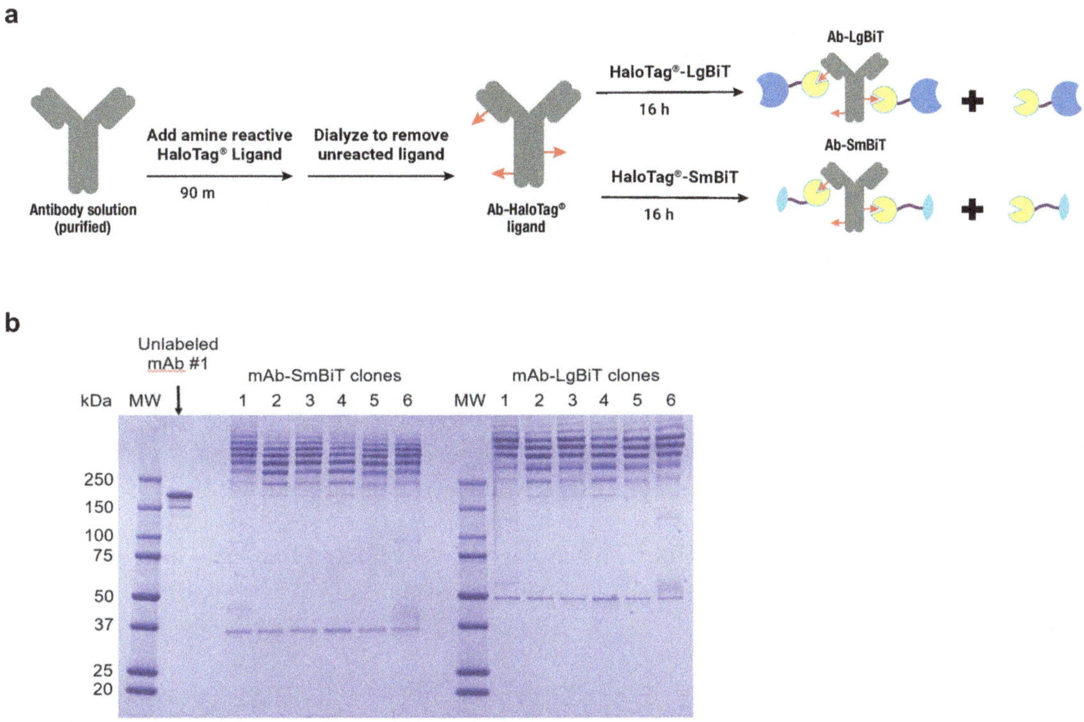

Fig. 2 Labeling antibodies with NanoBiT subunits. (**a**) Schematic of antibody labeling with HaloTag-LgBiT or HaloTag-SmBiT. A similar approach can be used to label other proteins. (**b**) Gel analysis of labeled antibodies for human IL-4 detection. Monoclonal antibodies (6) against human IL-4 were separately labeled with SmBiT and LgBiT. Subsequently, 2 μg antibody equivalents of each were resolved on a 4–20% Criterion TGX gel (Bio-Rad), then stained with Coomassie Blue to visualize the NanoBiT-labeled antibodies at molecular weights in excess of 150 kDa. Low molecular weight bands of ~33 and ~50 kDa are small amounts of residual HaloTag-SmBiT and HaloTag-LgBiT, respectively

antibody. The HaloTag component of HaloTag-NanoBiT makes a covalent bond with HaloTag ligand on the activated antibody and orients the BiT subunit.

3.1.1 Chemical Labeling of Antibodies with HaloTag Ligand

Reagents provided in Lumit Immunoassay labeling kit are sufficient to label 250 μg of antibody with HaloTag-LgBiT and 250 μg of antibody with HaloTag-SmBiT. At the recommended antibody concentration of 1.0 mg/mL, this amount corresponds to a volume of 250 μL, which is easy to handle during buffer exchange steps.

1. Buffer exchange antibody into 10 mM sodium bicarbonate (pH 8.5) using a desalting column (*see* **Notes 1** and **11**).
2. Add 20 molar excess of HaloTag ligand to the antibody solution. For example, add 1.7 μL of ligand to 250 μg of antibody (*see* **Note 12**).

3. If the antibody concentration is less than 1.0 mg/mL, add 50 molar excess of the ligand to the antibody sample. For example, add 4.25 μL of ligand to 250 μg of antibody (*see* **Note 13**).
4. Incubate at room temperature for 90 min with mixing.
5. Remove the unreacted HaloTag ligand using the Zeba desalting column, following the manufacturer's protocol.
6. Repeat the desalting process a second time for complete removal of HaloTag ligand (*see* **Note 14**).
7. Measure the concentration of HaloTag ligand-labeled antibody by measuring absorbance at 280 nm. The absorbance value at 280 nm (A_{280}) of a 1.0 mg/mL antibody solution is 1.4. Therefore, the antibody concentration in mg/ml can be determined by dividing the A_{280} reading by 1.4.

3.1.2 Conjugating HaloTag-SmBiT and HaloTag-LgBiT to Antibody-HaloTag Ligand

1. Add fourfold molar excess of HaloTag-LgBiT or HaloTag-SmBiT fusion protein to the antibody labeled with HaloTag ligand. For example, add 335 μL of fusion protein to 250 μg of antibody.
2. Add 5 μL of freshly prepared 10% IGEPAL CA-630 per 1 mL of solution from **step 1** to have a final IGEPAL CA-630 concentration of 0.05% (*see* **Note 15**).
3. Incubate for 16 h at 4 °C with gentle mixing.
4. Save a small aliquot for gel analysis.
5. If removing unreacted HaloTag-BiTs is desired, proceed to Subheading 3.1.3.
6. If Subheading 3.1.3 is not performed, we suggest adding glycerol to a final concentration of 50% to the labeled antibody solution from **step 3** and store at −20 °C in single-use aliquots.
7. Use the saved aliquot for non-denaturing SDS-PAGE gel to confirm efficient labeling (*see* **Note 7**). A representative gel image is shown in Fig. 2b.

3.1.3 Removing Unreacted HaloTag-BiTs Using Magne-HaloTag Beads

This step is optional. Unreacted HaloTag-SmBiT and HaloTag-LgBiT in solution do not interact with each other due to weak affinities (190 μM), and therefore do not cause significant background. If desired, the following protocol can be used to remove unreacted HaloTag-BiT.

1. Resuspend Magne HaloTag Beads by inverting the tube several times.
2. Transfer 85 μL of Magne HaloTag Beads to a 1.5-mL microcentrifuge tube. This amount is sufficient for the labeled antibody described in the previous section. Adjust the volume of Magne HaloTag Beads accordingly if the scale of antibody labeling is changed.

3. Place the tube on the magnetic stand for 30 s to capture the beads. Carefully remove the supernatant and discard.
4. Remove the tube from the magnetic stand and add 500 μL of PBS containing 0.05% IGEPAL CA-630.
5. Mix thoroughly for 2 min. Place the tube on the magnetic stand for 30 s. Carefully remove the supernatant and discard.
6. Repeat the wash two more times for a total of three washes.
7. After removing the third wash solution, add the antibody-BiT prepared in **step 3** of Subheading 3.1.2 to the equilibrated beads.
8. Incubate for 1 h at room temperature (22–25 °C) with constant mixing. Make sure the beads remain in suspension.
9. Place the tube on the magnetic stand for 30 s. Collect and save the supernatant, which contains the cleaned antibody-BiT.
10. Save a small aliquot for gel analysis.
11. We suggest adding glycerol to a final concentration of 50% to the labeled antibody solution from **step 9** and store at −20 °C in single-use aliquots.
12. Run a nonreducing SDS-PAGE gel to confirm the removal of unreacted HaloTag-BiT. Run control samples of initial components such as the antibody and BiTs as shown in Fig. 2b (*see* **Note 16**).

3.2 Lumit Cytokine Immunoassay

Cytokines are small proteins released from immune and nonimmune cells that signal through cell-surface receptors to exert their biological effects. They play critical inflammatory and immunomodulatory roles in the body and define an important class of drugs and drug targets for cancer and inflammatory disease [9]. As such, they are frequently measured in cell culture models for both basic research and drug discovery purposes, as well as for the characterization of cell therapies. Lumit technology delivers a simple and fast assay method for cytokine detection, requiring only a standard plate-reading luminometer for signal detection. Here, we describe how to build a Lumit Immunoassay with labeled primary antibodies using human IL-4 as the example target analyte.

3.2.1 Selection and Labeling of Candidate Primary Antibodies

The proper selection of candidate primary antibodies is an important first step in building a Lumit Immunoassay. While it is acceptable to include polyclonal antibodies, in the case of human IL-4, six monoclonal antibodies with high specificity for the target of interest were identified. There was no restriction on host species for the antibodies selected given that the intended application used only labeled primary antibodies with no reliance upon secondary antibodies. As required for labeling with SmBiT and LgBiT subunits, each of the selected antibodies were obtained as purified preparations (e.g., Protein A/G affinity purified) (*see* **Note 1**). The six candidate antibodies were each assigned an arbitrary screening I.D. (mAb1 through mAb6) for simplified tracking purposes (Table 1). As described in the previous section, the antibodies were labeled with NanoBiT subunits, cleaned from the majority of unbound HaloTag-NanoBiT fusion proteins, then analyzed by gel resolution and Coomassie Blue staining (Fig. 2b), as per guidance in Subheading 3.1. Labeled antibodies in PBS should be brought to 50% glycerol for long-term storage.

3.2.2 Rapid Screening of Candidate Antibody Pairs

Once each candidate antibody has been labeled separately with SmBiT or LgBiT subunits, all possible combinations of complementary NanoBiT-labeled antibodies are screened for their ability to detect the target analyte. A simple approach is to generate a combinatorial matrix (checkerboard) of SmBiT- and LgBiT-labeled antibodies in a 96-well plate as illustrated in Fig. 3 and test samples containing either a fixed concentration of analyte in culture medium to generate assay signal (S), or cell culture medium alone for background (B) determination. Following application of Lumit Detection Reagent B, determination of S/B ratios for all candidate antibody pairs indicates the optimal antibody pair(s) for target detection and further assay optimization. Of note, while in the IL-4 example provided here, similar performance was observed for specific antibody pairs (e.g., mAb5 and mAb6) independent of the configuration of NanoBiT labeling, moderate or even dramatic performance differences can occur for certain targets based on

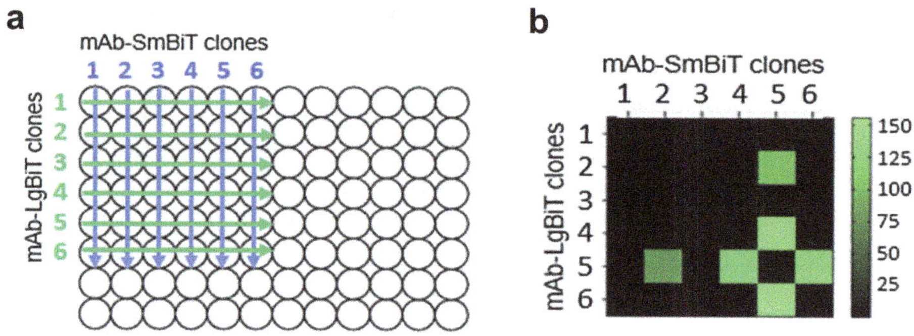

Fig. 3 Rapid screening of candidate antibody pairs for human IL-4 detection. Each SmBiT-labeled mAb was combined with each LgBiT-labeled mAb and assayed for possible detection of human IL-4. (**a**) mAb-SmBiT clones 1–6 were dispensed into columns 1–6, respectively, while mAb-LgBiT clones 1–6 were dispensed into rows A–F, respectively, of the same 96-well plate to produce 2× working concentrations of each antibody pair. The 2× working solutions of each antibody pair were transferred into corresponding wells of two 96-well plates containing an equal volume of 1 ng/mL recombinant human IL-4 for signal determination. Likewise, 2× antibody combinations were transferred into two 96-well plates containing culture medium, but no cytokine, for background determinations. (**b**) Average signal (S) and background (B) determinations were used to calculate S/B ratios, subsequently represented in heatmap fashion. Note that the final screening antibody concentrations were 250 ng/mL

which antibody in a pair is conjugated with SmBiT versus LgBiT. The combinatorial matrix approach outlined in this section efficiently tests all possible productive antibody pairings and configurations.

1. Prepare a 4× dilution (1 μg/mL) of each labeled antibody in culture medium (e.g., RPMI 1640, 10% FBS) (*see* **Note 17**). Total volume of each 4× antibody dilution should be more than sufficient for application below (e.g., 850 μL 4× dilution of each labeled, candidate anti-hIL-4 antibody).
2. Prepare a master plate containing a combinatorial matrix of candidate antibody pairs as described in Fig. 3a.
   (a) Dispense 125 μL of 4× SmBiT-labeled antibodies into each well of assigned columns of a 96-well plate.
   (b) Dispense 125 μL of 4× LgBiT-labeled antibodies into assigned rows of the same 96-well plate to generate 2× mixtures of the various antibody pairs (Fig. 3a).
3. Briefly mix the 2× antibody combinations using an orbital plate shaker (10–15 s, 250–350 rpm).
4. Prepare 4 mL of 1 ng/mL solution (or any relevant fixed concentration) of recombinant human IL-4 (hIL-4) in culture medium.
5. In two separate 96-well white plates, dispense 50 μL of 1 ng/mL hIL-4 into positions A1 through F6. These plates will be used for duplicate signal (S) determinations for each antibody pair (*see* **Note 18**).
6. In two additional 96-well, white plates, dispense 50 μL of culture medium into positions A1 through F6. These plates will be used for duplicate background (B) determinations for each antibody pair (*see* **Note 18**).
7. Replica transfer 50 μL of the 2× mixtures for each antibody pair into corresponding well positions in the two plates containing 1 ng/mL hIL-4 as well as the two plates containing only culture medium. Final antibody concentrations equal 250 ng/mL.
8. Briefly mix using an orbital plate shaker (10–15 s, 350–450 rpm).
9. Incubate for 60–90 min at room temperature.
10. Prepare sufficient volume of room temperature Lumit Detection Reagent B by performing a 1:20 dilution of Lumit™ Detection Substrate B into room temperature Lumit™ Detection Buffer B (*see* **Note 19**).
11. Add 25 μL of Lumit™ Detection Reagent B into each well of the two assay plates containing 1 ng/mL hIL-4 for signal (S) determination, as well as the two assay plates without analyte for background (B) determination.

12. Briefly mix using an orbital plate shaker (10–15 s, 350–450 rpm).
13. Incubate for 3–5 min at room temperature.
14. Read luminescence (*see* **Note 20**).
15. Average duplicate signal and background luminescence determinations, then calculate the S/B ratio for each candidate antibody pair (Fig. 3b).

3.2.3 Lumit Cytokine Immunoassay Optimization

Following completion of the rapid screen, candidate antibody pairs are rank ordered based on observed S/B ratios for detection of a fixed analyte concentration. At that point, one (or more) antibody pair can be selected for further assay development. In the case of human IL-4 assay development, the pair of mAb5-SmBiT and mAb6-LgBiT delivered an attractive S/B ratio (although other pairings were also worthy of consideration). To optimize assay performance, a range of concentrations of mAb5-SmBiT and mAb6-LgBiT were tested against one another in the presence and absence, again, of a fixed concentration of analyte. S/B ratios were assessed to establish the respective concentration of each antibody in the pair to deliver optimal assay performance.

1. Prepare a range of 4× dilutions of each labeled antibody in culture medium (RPMI 1640, 10% FBS) for the selected antibody pair (*see* **Note 17**). For mAb5-SmBiT and mAb6-LgBiT, the chosen 4× dilutions for each antibody were 0.2, 0.4, 0.6, 0.8, 1.0, 1.2, 1.4, and 1.6 μg/mL (producing final in-assay concentrations of 50, 100, 150, 200, 250, 300, 350, and 400 ng/mL of each antibody). The total volume of each 4× antibody dilution should be more than sufficient for application below (e.g., 1100 μL of each 4× dilution for mAb5-SmBiT and mAb6-LgBiT).
2. Prepare a master plate containing all combinations of the various dilutions for each labeled antibody in the selected pair to be optimized as described in Fig. 4.
   (a) Dispense 125 μL of mAb5-SmBiT 4× dilutions into each well of assigned columns of a 96-well plate.
   (b) Dispense 125 μL of mAb6-LgBiT 4× dilutions into assigned rows of the same 96-well plate to generate 2× dilutions of each pairing of antibody concentrations.
3. Briefly mix 2× antibody dilutions using an orbital plate shaker (10–15 s, 250–350 rpm).
4. Prepare 7 mL of a desired concentration of recombinant human IL-4 (hIL-4) in culture medium. Here, 10 ng/ml IL-4 solution was made.

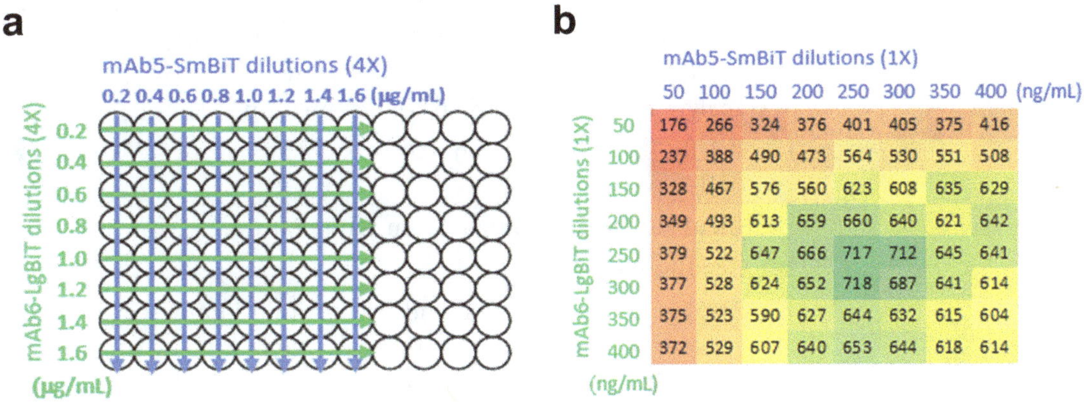

Fig. 4 Optimization of selected antibody concentrations. The concentrations of labeled antibodies in the selected pair from rapid screening (mAb5-SmBiT and mAb6-LgBiT) (Fig. 3) were optimized for detection of human IL-4. (**a**) A range of mAb5-SmBiT concentrations were dispensed into columns 1–8, while identical concentrations of mAb6-LgBiT were dispensed into rows A–H, respectively, of the same 96-well plate to produce 2× concentrations of the antibody pair. The 2× working solutions of the antibody pair were transferred into corresponding wells of two 96-well plates containing an equal volume of 10 ng/mL recombinant human IL-4 for signal determination. Likewise, the 2× antibody combinations were transferred into two 96-well plates containing culture medium, but no cytokine, for background determinations. (**b**) Average signal (S) and background (B) determinations were used to calculate S/B ratios, subsequently represented in heatmap fashion. Optimal assay S/B was determined to be achieved at approximately 275 ng/mL of each labeled antibody

5. Dispense 50 µL of 10 ng/mL IL-4 into positions A1 through H8 of two separate 96-well, white plates. These plates will be used for duplicate signal (S) determinations for each pairing of antibody concentrations.
6. Dispense 50 µL of culture medium into positions A1 through H8 of two additional 96-well, white plates. These plates will be used for duplicate background (B) determinations for each pairing of antibody concentrations.
7. Replica transfer 50 µL of the 2× antibody dilutions into corresponding well positions in the two plates containing 10 ng/mL hIL-4 as well as the two plates containing only culture medium.
8. Briefly mix using an orbital plate shaker (10–15 s, 350–450 rpm).
9. Incubate for 60–90 min at room temperature.
10. Prepare sufficient volume of room temperature Lumit™ Detection Reagent B by performing a 1:20 dilution of Lumit™ Detection Substrate B into room temperature Lumit™ Detection Buffer B (*see* **Note 19**).

11. Add 25 μL of Lumit™ Detection Reagent B into each well of the two assay plates containing 10 ng/mL hIL-4 for signal (S) determination, as well as the two assay plates without analyte for background (B) determination.
12. Briefly mix using an orbital plate shaker (10–15 s, 350–450 rpm).
13. Incubate for 3–5 min at room temperature.
14. Read luminescence.
15. Average duplicate signal and background luminescence determinations, then calculate the S/B ratio for each pairing of antibody concentrations (Fig. 4b).

3.2.4 Calibration Curve and Cell-Based Detection of IL-4

After determining the optimal concentration of each antibody in the selected pair for assay development, it is necessary to assess assay performance against a calibration range of recombinant analyte concentrations as well as against cell-based samples. The resultant standard curve should sufficiently cover the required range of analyte concentrations, keeping in mind that excessive analyte concentrations can produce a "hook" effect. A desirable standard curve should reflect the required sensitivity of the intended assay application as well as a broad range of relative linearity, a beneficial Lumit feature that mitigates the need for sample dilutions. Cell-based performance can be evaluated by direct reagent addition to treated cell wells or, alternatively, to sampled medium transferred to a separate assay plate. The following is a description of the completely homogeneous, "direct" assay protocol applied to both recombinant standard (Fig. 5a) and an appropriate cell-based model (Th2 cells) for human IL-4 release (Fig. 5b).

Cell Plating and Treatment

1. Plate cells into a 96-well, white tissue culture plate (*see* **Note 21**). Allow cells to attach (typically overnight) if using adherent cells. Leave columns 1 and 2 empty for the standard dilutions to be prepared in the next section.
2. Treat cells by adding a volume of test agent (e.g., Cell Stimulation Cocktail) to each well such that the resultant volume is 80 μL per well for a 96-well plate (*see* **Note 22**). For example, if 60 μL of cells are plated per well, add 20 μL of 4× treatment agent in culture medium.

Recombinant Human IL-4 Standard Dilutions

1. Thaw human IL-4 standard (10 μg/mL in PBS with 0.1% w/v BSA) at room temperature shortly before use (*see* **Notes 23** and **24**). Finger flick thawed standard and quick spin before opening.

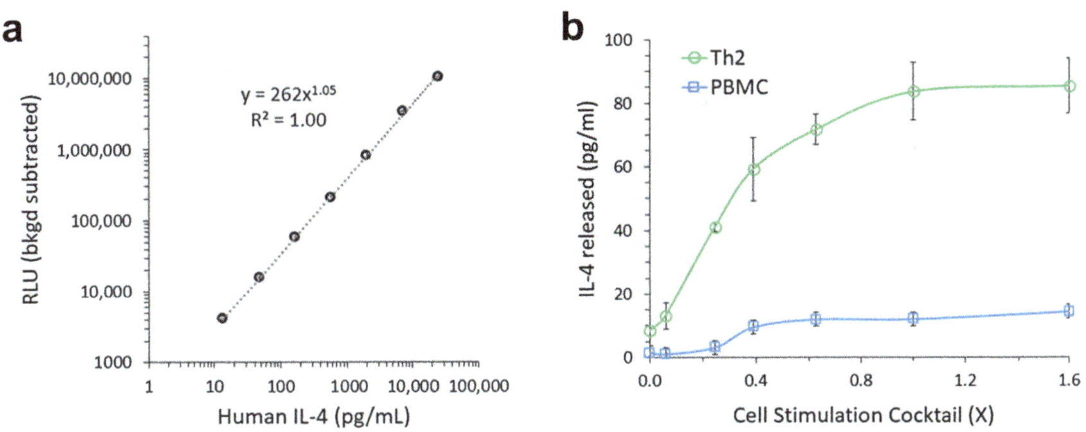

Fig. 5 Calibration curve and cell-based application of Lumit hIL-4 Immunoassay. The optimized pair of labeled antibodies for human IL-4 detection (mAb5-SmBiT and mAb6-LgBiT) was applied for the detection of IL-4 released from human cell culture models. (**a**) A calibration curve for Lumit Immunoassay detection of recombinant human IL-4 standard dilutions (13.6–25,000 pg/mL) was generated in the same 96-well plate used for assessment of cell-based IL-4 release. (**b**) 30,000 human T helper 2 (Th2) cells or 30,000 human peripheral blood mononuclear cells (PBMCs), respectively, were plated into an all-white 96-well plate, allowed to recover overnight, then treated for 24 h with increasing concentrations of Cell Stimulation Cocktail (eBioscience). Subsequently, the Lumit hIL-4 Immunoassay assay reagents were added directly to the treated cell wells (and analyte standards) and luminescent signals obtained were interpolated against the calibration curve using cubic spline fit to determine the dose-dependent release of IL-4 from each cell model. Results obtained are the average of triplicate determinations. As expected, clear evidence of dose-dependent release of IL-4 was observed in Th2 cells, while IL-4 release from PBMC was negligible

2. Dilute human IL-4 standard to 25,000 pg/mL by adding 2 μL of the provided standard to 798 μL of culture medium.
3. Set up seven tubes (or seven chambers of a dilution reservoir) with 250 μL culture medium in each tube (or chamber).
4. Prepare 3.5-fold serial dilutions of the standard (dilution factor previously optimized) by transferring 100 μL from the 25,000 pg/mL IL-4 dilution to 250 μL of culture medium, pipetting up and down to mix.
5. Repeat five more times to generate seven standard dilutions with a range of 13.6–25,000 pg/mL (*see* **Note 25**). The last tube (or chamber) should contain only culture medium as the background control.
6. After the cell treatment is complete, add 80 μL of the standard dilutions and zero-standard background control to duplicate wells in empty columns 1 and 2 of the 96-well, white assay plate.

Lumit IL-4 Immunoassay Protocol

1. Remove 500× anti-hIL-4 antibody stocks (mAb5-SmBiT and mAb6-LgBiT) from −20 °C immediately before use (*see* **Notes 17** and **26**).

2. Finger flick and quick spin the antibody vials before opening.
3. Immediately prior to use prepare a 5× antibody mixture by performing a 1:100 dilution of both antibodies into a single volume of culture medium, followed by pipette mixing of the resultant solution. (For example, 24 μL mAb5-SmBiT +24 μL mAb6-LgBiT +2.4 mL culture medium.)
4. Add 20 μL of the 5× antibody mixture to the assay wells in the 96-well, white plate containing treated cells and standards.
5. Briefly mix using an orbital plate shaker (10–15 s, 250–350 rpm).
6. Incubate for 60–90 min at 37 °C in a 5% CO_2 humidified incubator (*see* **Note 27**).
7. Prepare sufficient volume of room temperature Lumit™ Detection Reagent B by performing a 1:20 dilution of Lumit™ Detection Substrate B into room temperature Lumit™ Detection Buffer B (*see* **Note 19**).
8. Equilibrate the assay plate to room temperature for 10–15 min.
9. Add 25 μL of Lumit™ Detection Reagent B to each assay well of the 96-well plate.
10. Briefly mix using an orbital plate shaker (10–15 s, 300–500 rpm).
11. Incubate for 3–5 min at room temperature.
12. Read luminescence (*see* **Note 20**).
13. Interpolate luminescence signal from cell wells against the generated calibration curve (Fig. 5a) to determine levels of released human IL-4 from the cell models (Fig. 5b).

3.3 Lumit Phospho-ERK1 (Thr202) Immunoassay

Lumit Immunoassay technology can be used to detect protein phosphorylation in a cell-based assay. For this example, we targeted phospho-ERK (pERK). Since sensitivity and specificity of the Lumit Immunoassay depend on the quality of primary antibodies used, it is important to select antibodies that perform well in a Lumit Immunoassay as opposed to other applications. To select the optimal antibody pair for pERK (Thr202 site) detection, a checkerboard experiment was performed following the manufacturer's instructions [10] using multiple antibody pairs at a single antibody concentration (150 ng/mL) with cells treated to provide high- and low-end signals. Using this approach, we selected an antibody pair that provides the highest luminescent signal with the largest signal-to-background ratio (*see* **Note 28**). Then, a second checkerboard experiment was performed with titrations of the two antibodies of the selected pair to identify optimal antibody concentration. Upon activation of the MAPK pathway with stimulants such as EGF, ERK1 is phosphorylated in MCF-7 cells. After lysis of the cell

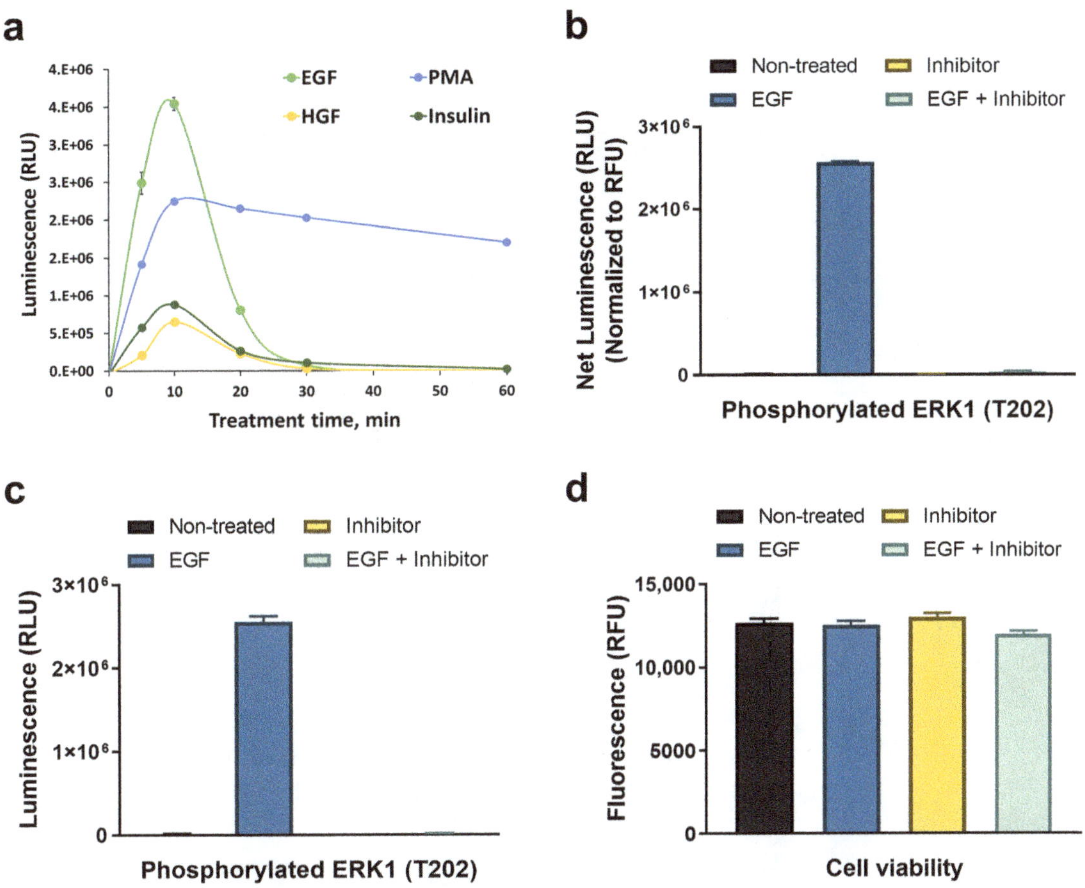

Fig. 6 Detection of phosphorylated ERK1 in MCF-7 cell lysate. (**a**) Phosphorylation of ERK1 in response to different stimulants at different treatment times. 50,000 seeded MCF-7 cells were starved overnight. The cells were then untreated or treated with EGF (100 ng/mL), PMA (1 μM), HGF (100 ng/mL), or insulin (1 μM) for the indicated times. (**b–d**) Inhibitory effect of a MEK kinase inhibitor on phosphorylation of ERK1. 50,000 seeded MCF-7 cells were starved overnight. The cells were then left untreated or pretreated with U0126 (10 μM, 1 h) before they were untreated or treated with EGF (100 ng/mL) for 5 min. Phosphorylated ERK1 levels were measured using Lumit Immunoassay Cellular System. (**b**) Normalized luminescence values calculated as described in **Note 33**. (**c**) Net luminescence values. (**d**) Net fluorescence values

membrane, phospho-ERK1 (Thr202) can be detected using the Lumit Immunoassay Cellular System in combination with the selected anti-ERK1 and anti-pERK1 antibodies (Fig. 6).

3.3.1 Detection of Phosphorylated ERK1 in MCF-7 Cells

Following Table 2 scheme, treat cells with the indicated MAPK signaling stimulants at different times to generate a time course of ERK1 phosphorylation (Fig. 6a).

1. Maintain MCF-7 cells in complete growth medium at 37 °C in 5% CO_2. Passage cells at 70–90% cell confluency.
2. Count cells and dispense 50,000 cells in 160 μL per well of a whole 96-well plate in the complete growth medium (*see* **Note 29**).

Table 2
Plate arrangement for ERK1 phosphorylation detection in response to different stimulants

	1	2	3	4	5	6	7	8	9	10	11	12	
EGF (min)	0	5	10	20	30	60	0	5	10	20	30	60	PMA (min)
A			No primary Ab control						No primary Ab control				
B													
C			Phospho-ERK (T202)						Phospho-ERK (T202)				
D													
E													
F			Phospho-ERK (T202)						Phospho-ERK (T202)				
G													
H			No primary Ab control						No primary Ab control				
HGF (min)	0	5	10	20	30	60	0	5	10	20	30	60	Insulin (min)

3. Incubate the plate at 37 °C in 5% CO_2 overnight.
4. Remove the culture medium carefully using a multichannel pipette.
5. Add 100 μL of starvation medium (*see* **Note 30**).
6. Incubate the plate at 37 °C in 5% CO_2 overnight.
7. Prepare various stimulation media. Here, we used starvation medium containing 100 ng/mL EGF, 1 μM PMA, 100 ng/mL HGF, or 1 μM insulin.
8. To start the time course of stimulation, treat the cells for the longest time point of 60 min first. Remove the starvation medium from columns 6 and 12 carefully using a multichannel pipette as shown in Table 2.
9. Add 40 μL of each stimulant medium to their corresponding wells in columns 6 and 12 (e.g., EGF in wells A6/B6/C6D6 and insulin in wells E12/F12/G12/H12).
10. Incubate the plate at 37 °C in 5% CO_2 for 30 min.
11. Remove the starvation medium from columns 5 and 11 carefully using a multichannel pipette.
12. Add 40 μL of each stimulant to their corresponding wells in columns 5 and 11 to start treatment for the 30 minutes time point.
13. Incubate the plate at 37 °C in 5% CO_2 for 10 min.
14. Repeat these steps for each time point by incubating the plate with stimulants for the needed time.
15. At time 0, remove the starvation medium from columns 1 and 7 carefully and add 40 μL of the starvation medium for the 0 min time point.
16. Proceed to cell lysis to start the Lumit Cellular Immunoassay described in Subheading 3.3.3.

Table 3
Plate arrangement for ERK1 phosphorylation detection in response to MEK1 inhibitor

<table>
<tr><th></th><th>1</th><th>2</th><th>3</th><th>4</th><th>5</th></tr>
<tr><td></td><td>No treatment</td><td>EGF</td><td>U0126</td><td>EGF + U0126</td><td></td></tr>
<tr><td>A</td><td colspan="4" rowspan="3">Phospho-ERK (T202)</td><td></td></tr>
<tr><td>B</td><td></td></tr>
<tr><td>C</td><td></td></tr>
<tr><td>D</td><td colspan="4" rowspan="3">No primary Ab control</td><td></td></tr>
<tr><td>E</td><td></td></tr>
<tr><td>F</td><td></td></tr>
<tr><td>G</td><td></td><td></td><td></td><td></td><td></td></tr>
<tr><td>H</td><td></td><td></td><td></td><td></td><td></td></tr>
</table>

3.3.2 Testing Effect of a MEK Kinase Inhibitor on Phosphorylation of ERK1

Upon activation of MAPK pathway with EGF, MEK kinase phosphorylates ERK1 in MCF-7 cells. MEK kinase inhibitors such as U0126 inhibit this phosphorylation. Following Table 3 plate scheme, treat cells with EGF in the presence or absence of U0126 and then measure ERK1 phosphorylation with the Lumit Immunoassay Cellular System (Fig. 6b). To determine whether U0126 affects cell viability and normalize data to viable cell number, add GF-AFC cell viability substrate (fluorogenic, cell-permeant, live cell substrate) to the cells 60 min before lysis. At the end of the immunoassay, record luminescence (RLU) and fluorescence (RFU) in a multimode plate reader and normalize RLUs to RFUs (*see* Fig. 6c, d and **Note 31**).

1. Maintain MCF-7 cells in complete growth medium at 37 °C in 5% CO_2. Passage cells at 70–90% cell confluency.
2. Count cells and dispense 50,000 cells in 160 μL per well of a whole 96-well plate in the complete growth medium (*see* **Note 29**).
3. Incubate the plate at 37 °C in 5% CO_2 overnight.
4. Remove the culture medium carefully using a multichannel pipette (*see* **Note 30**).
5. Add 100 μL of starvation medium.
6. Incubate the plate at 37 °C in 5% CO_2 overnight.
7. Prepare media containing GF-AFC and U0126 inhibitor or DMSO.
8. Remove the media carefully using a multichannel pipette.

9. Add 35 μL of the GF-AFC/U0126 medium to wells indicated with U0126 and 35 μL GF-AFC/DMSO medium to the wells indicating no treatment or EGF treatment only as shown in Table 3.
10. Incubate the plate at 37 °C in 5% CO_2 for 1 h.
11. Prepare EGF stimulation medium 2. Here, we used starvation medium containing 800 ng/mL EGF.
12. Add 5 μL of EGF stimulation medium 2 to the wells indicating EGF treatment and 5 μL starvation medium to the rest of the wells as indicated in Table 3.
13. Incubate the plate at 37 °C in 5% CO_2 for 5 min.
14. Proceed to cell lysis to start the Lumit Cellular Immunoassay described in Subheading 3.3.3.

3.3.3 Lumit Immunoassay Cellular System Protocol

1. Treat cells as described in Subheading 3.3.1 or 3.3.2.
2. Add 10 μL lysis solution to all wells containing cells in 40 μL medium.
3. Shake the plate vigorously at around 800 rpm for 40 min.
4. Prepare the Lumit p-ERK antibody mix containing the anti-ERK/anti-pERK antibody pair and the Lumit secondary antibody pair and the "Lumit reagent control" mix containing only the Lumit secondary antibody pair.
5. Add 50 μL of each of the antibody mixes to their corresponding wells as indicated in Table 2 or 3.
6. Shake the plate gently at around 400 rpm for 2 min.
7. Incubate the plate at 23 °C for 90 min (*see* **Note 32**).
8. Prepare Lumit Detection Reagent (*see* **Note 5**).
9. Add 25 μL Lumit Detection Reagent to all the wells.
10. Shake the plate gently at around 400 rpm for 2 min.
11. Sequentially measure luminescence and fluorescence with a multimode plate reader.
12. To analyze the data, subtract the "Lumit reagent" control RLU from the RLU values of each of the treatments to obtain net luminescence values. If GF-AFC cell viability substrate is used, normalize RLU values to RFU values (*see* **Note 33**).

3.4 Lumit SARS-CoV2 Spike RBD: hACE2 Immunoassay

Lumit immunoassays can be developed to detect PPIs by using NanoBiT-labeled antibodies against the tags of interacting recombinant proteins. Here, we describe the detection of the PPI between viral SARS-CoV2 spike protein receptor-binding domain (RBD) and human ACE2 recombinant proteins tagged with rabbit and mouse Fc domains, respectively. To detect the interaction, a rabbit Fc-domain-tagged RBD and a mouse Fc-domain-tagged

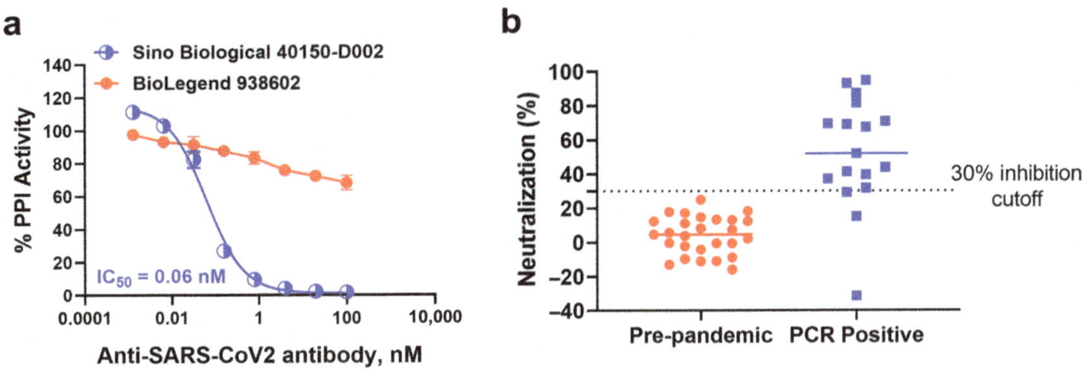

Fig. 7 SARS-CoV-2 spike RBD: human ACE2 PPI immunoassay. Interaction between the rabbit Fc-tagged RBD fragment of SARS-CoV-2 spike protein and the soluble mouse Fc-tagged human ACE2 receptor is detected by LgBiT/SmBiT-conjugated Lumit secondary antibodies, generating luminescent signal. (**a**) Dose-dependent inhibition of SARS-CoV-2 spike RBD:ACE2 interaction using two monoclonal antibodies that recognize RBD with different neutralizing potencies. (**b**) Scatter plot of patient-derived negative pre-pandemic and COVID-19 patient serum samples. Negative pre-pandemic samples were used to establish the cutoff values (i.e., 30%) to enable the identification of patient-derived samples with different neutralizing antibody levels

ACE2 are mixed with SmBiT- and LgBiT-conjugated secondary antibodies (*see* **Note 34**). Interaction of RBD and ACE2 along with binding of the NanoBiT-labeled secondary antibodies to their corresponding epitopes on the Fc-domains brings NanoBiT subunits into proximity to generate light in proportion to the amount of PPI (Fig. 1c). This immunoassay is suitable for high-throughput screening and detection of not only small or large molecules interfering with the RBD:ACE2 PPI (Fig. 7a) but also can be used as a surrogate assay for detecting neutralizing antibodies in buffer, COVID-19 patient serum, or plasma samples (Fig. 7b).

3.4.1 General Protocol of Lumit SARS-CoV 2 Spike RBD: hACE2 PPI Immunoassay

This is a general protocol for Lumit PPI immunoassay used to detect PPI in the presence of potential inhibitors of SARS-CoV2 spike RBD: ACE2 PPI. Specific protocols for antibody titration or detection of neutralizing effect of COVID-19 patient serums are described in the next section.

1. Add 5 μL sample (50% patient-derived serum or plasma, neutralizing antibody, or compound).
2. Add 10 μL of 7.5 nM rFc-RBD solution.
3. Add 10 μL of 7.5 nM mFc-ACE2 solution.
4. Add 25 μL Lumit secondary antibody mix.
5. Incubate 60 min at room temperature.
6. Add 12.5 μL Lumit Detection Reagent.
7. Incubate 30 min at room temperature.
8. Read luminescence.

3.4.2 Detection of Neutralizing Antibody Interference with RBD: ACE2 PPI

The following protocol established for titrating a neutralizing anti-SARS-CoV2 antibody is formatted to a regular 96-well assay plate. Lower volumes and different plate formats may be used, provided that a 2:2:1 volume ratio is followed [i.e., RBD/ACE2 (25 μL): Lumit Antibody Mix (25 μL): Lumit Detection Reagent (12.5 μL)].

1. Dispense 20 μL of 1× immunoassay reaction buffer into wells A2 to A12 of a 96-well white preparative plate.
2. Prepare 40 μL of 5 μM (0.75 mg/mL) anti-SARS-CoV-2 spike antibody solution by mixing 30 μL of 1 mg/mL antibody with 10 μL of 1× immunoassay reaction buffer.
3. Dispense the 5 μM antibody solution to well A1. Serially dilute antibody by transferring 20 μL from well A1 to A2 and mix by pipetting 6–8 times. Continue the serial dilution by transferring 20 μL from well A2 to A3, and from A3 through A10. Control well A11 is for positive controls (i.e., no inhibitory antibody) and A12 is reserved for Lumit Reagent Only control (only Lumit secondary antibodies). The antibody dilution curve is ready for transfer to assay plate.
4. Transfer 5 μL of the inhibitory antibody dilutions in duplicates to a 96-well white assay plate by transferring dilution series from row A (inhibitor preparative plate) to rows A and B (assay plate). The antibody curve concentration will be diluted 10-fold in the assay reaction (i.e., 50 μL).
5. Dispense 10 μL of 7.5 nM RBD-rFc solution into wells A1 to A11 and B1 to B11.
6. Transfer 10 μL of 1× immunoassay reaction buffer into wells A12 and B12 (Lumit Reagents Only Control). Shake the assay plate for 2 min and briefly centrifuge it at 1000 rpm.
7. Incubate plate for 30 min at 23 °C.
8. Dispense 10 μL of 7.5 nM ACE2-mFc solution into wells A1 to A11 and B1 to B11.
9. Transfer 10 μL of 1× immunoassay reaction buffer into wells A12 and B12 (Lumit Reagents Only Control).
10. Dispense 25 μL of Lumit Antibody Mix into all wells (A1 to B12). Briefly centrifuge assay plate. Shake plate for 2 min.
11. Incubate plate for 60 min at 23 °C.
12. Prepare Lumit Detection Reagent as described previously.
13. Dispense 12.5 μL of Lumit Detection Reagent into all wells (A1 to B12). Briefly centrifuge assay plate. Shake plate for 2 min.
14. Incubate plate for 30 min at 23 °C.
15. Measure luminescence in a plate-reading luminometer using an integration time of 0.5 s per well.

3.4.3 Evaluation of Patient-Derived Samples for RBD: ACE2 PPI Neutralizing Effect

The following protocol is used with different types of patient-derived samples (i.e., serum, plasma) for detecting anti-SARS-CoV-2 spike RBD neutralizing antibodies (nAbs). We recommend performing these assays in a 96-well white assay plate.

1. Transfer 5 μL 50% diluted patient-derived sample directly into the 96-well assay plate. The final patient-derived sample concentration will become 5% after the addition of the remaining reagents (*see* **Notes 9** and **10**).
2. Use 5–10 PCR-verified pre-pandemic negative COVID-19 samples as "Pre-COVID19 Negative Sample Controls" (i.e., negative samples + spike RBD: ACE2) and background samples as "Lumit Reagents Only Controls" (i.e., negative samples without spike RBD: ACE2).
3. Dispense 5 μL of 50% diluted pre-COVID-19-negative samples in all designated control wells ("Pre-COVID19-Negative Sample Controls" and "Lumit Reagents Only Controls").
4. Dispense 10 μL of 7.5 nM RBD-rFc solution into wells containing patient-derived samples and Pre-COVID-19-Negative Sample Controls.
5. Dispense 10 μL of 1× immunoassay reaction buffer into Lumit Reagents Only Control wells. Shake the assay plate for 2 min and briefly centrifuge it at 500–1000 rpm.
6. Incubate plate for 30 min at 23 °C.
7. Dispense 10 μL of 7.5 nM ACE2-mFc solution into patient-derived sample wells and into Pre-COVID19-Negative Sample Controls.
8. Transfer 10 μL of 1× immunoassay reaction buffer into Lumit Reagents Only Control wells.
9. Dispense 25 μL of Lumit Antibody Mix into all wells. Shake the assay plate for 2 min and briefly centrifuge it at 1000 rpm.
10. Incubate plate for 60 min at 23 °C.
11. Dispense 12.5 μL of Lumit Detection Reagent into all wells. Shake plate for 2 min.
12. Incubate plate for 30 min at 23 °C.
13. Measure luminescence in a plate-reading luminometer using an integration time of 0.5 s per well.

3.4.4 Lumit PPI Data Analysis

Data Analysis for Antibody Titration

1. Subtract the "Lumit Reagents Only" relative luminescence unit values (RLUs) from the RLUs across the plate to obtain net luminescence values.
2. Transform net luminescence values to percent activity values by using the following formula:

$$\text{Activity (\%)} = \left(\text{Net Luminescence}_{(\text{sample})} / \text{Net Luminescence}_{(\text{Average of “Pre-COVID19 negative serum” controls})}\right) \times 100$$

3. Graph the percent activity data versus antibody inhibitor concentration and perform curve fitting using nonlinear regression analysis to determine the IC_{50} value (Fig. 7a).

Data Analysis for Patient-Derived Sample Evaluation

1. Subtract the “Lumit Reagents Only” RLUs from the RLUs across the plate to obtain net luminescence values.
2. Transform net luminescence values to percent PPI activity by using the following formula:

$$\text{Activity (\%)} = \left(\text{Net Luminescence}_{(\text{sample})} / \text{Net Luminescence}_{(\text{Average of “Pre-COVID19 negative serum” controls})}\right) \times 100$$

The “Lumit Reagents Only” and “Pre-COVID19 Negative Serum” controls will be 0 and 100%, respectively.

3. Transform percent activity values to percent neutralization values:

$$\text{Neutralization (\%)} = 100 - \text{Activity (\%)}$$

4. Graph the percent neutralization data in a scatter or box plot (Fig. 7b).
5. A cutoff value of 30% neutralization should be sufficient to indicate if a serum is positive or negative for the presence of neutralizing antibodies. Values above 30% neutralization are considered positive.

4 Notes

1. Antibody needs to be free of any protein stabilizers like BSA or amine-containing buffers or additives like Tris, glycine, sodium azide, etc.
2. Primary, human T helper 2 (Th2) cells were prepared via differentiation of purified, human CD4+ T cells.
3. Store Lumit™ Anti-Mouse Ab-LgBiT, Lumit™ Anti-Rabbit Ab-SmBiT, Lumit™ Detection Substrate C on ice before use.
4. These volumes can be adjusted up and down depending on the number of wells being used in each experiment.
5. Lumit Antibody Mix and Lumit Detection Reagent are prepared just prior to use.

6. Concentrated stock of RBD and ACE2 can be stored in this buffer at −80 °C.
7. Store SARS-CoV-2 RBD (rabbit Fc), 0.5 μM, human ACE2 (mouse Fc), 0.5 μM, Lumit™ Anti-Mouse Ab-LgBiT, Lumit™ Anti-Rabbit Ab-SmBiT, Lumit™ Detection Substrate C on ice before use.
8. The volumes of the solutions made here are enough for 50 assays. These volumes can be adjusted up and down depending on the number of wells being used in each experiment.
9. Collect patient-derived samples using your institution standard procedures. Follow the instructions provided with your collection device for use and processing of the sample.
10. Samples can be tested immediately after collection or stored at 2–8 °C for up to 3 days. Freeze samples at or below −20 °C for longer storage times.
11. The optimum pH for labeling the primary amine of lysine is pH 8.5. We recommend 10–100 mM sodium bicarbonate buffer (pH 8.5) for labeling. Proteins in other buffer solutions may be buffer-exchanged to adjust the pH. If proteins are in PBS (pH 7.2), add 1/10th the volume of 1 M bicarbonate buffer at pH 8.5 to adjust the pH.
12. At 1.0 mg/mL antibody, it is recommended to add 1.7 μL of ligand to 250 μg of antibody. If only 100 μg antibody is used, add 0.7 μL of ligand.
13. For optimum labeling, the antibody concentration should be 1.0 mg/mL or higher. Labeling efficiency drops if the antibody concentration is below 1.0 mg/mL and can be improved by adding higher amount of amine-reactive ligand (4.25 or 1.7 μL of ligand to 250 or 100 μg of antibody, respectively).
14. Unreacted HaloTag ligand will bind to the HaloTag-NanoBiT and block it from labeling antibody.
15. Adding IGEPAL CA-630 can minimize occasional precipitation of NanoBiT-labeled antibodies. IGEPAL CA-630 solution should be prepared fresh.
16. Successful labeling of an antibody is determined by disappearance of unlabeled antibody at 150 kDa and appearance of multiple high molecular weight bands corresponding to antibody conjugated with HaloTag-NanoBiTs (Fig. 2b).
17. Remove Lumit™ Buffer B from −20 °C at the same time and equilibrate to room temperature if not already thawed.
18. To minimize the number of plates used, the duplicate signals for hIL4 (S) and culture medium (B) can be generated by using

wells A7 through F12 of each plate similarly to A1 through F6 design.

19. Once reconstituted, the Lumit™ Detection Reagent B will lose 10% activity in approximately 3 h at 20 °C. At 4 °C, the reconstituted reagent will lose 10% activity in approximately 7 h.
20. Assay signal is stable with a half-life of approximately 2 h, compatible with batch processing of multiple assay plates. It is recommended that standard controls are included in each plate.
21. While the broad dynamic range of the assay offers considerable flexibility for cell number, the optimal number of cells dispensed per well for a specific cell model should be empirically determined. Ensure the maximum level of cytokine released does not exceed the linear range of the detection chemistry. Within that constraint, cell number can be increased to meet the detection requirements of low-level cytokine production.
22. Assays can be performed in 384-well, white plates by scaling down the cell number and volumes of all assay components.
23. Prepare serial human IL-4 standard dilutions shortly before cell treatment incubations are completed.
24. Extra human IL-4 standard (10 μg/mL) can be dispensed into aliquots and stored at −20 °C. Avoid multiple freeze-thaw cycles.
25. Change pipette tips between each dilution step to avoid analyte carryover. The linear range of the assay is large, so carryover from high to low concentrations can compromise the standard curve.
26. Stock concentrations of mAb5-SmBiT and mAb6-LgBiT were adjusted to 500× of the "optimized" final assay concentrations (275 ng/mL of each) prior to use.
27. While cells are incubating with labeled antibodies, prepare Lumit™ Detection Reagent B.
28. During antibody selection, both set 1 and set 2 of the Lumit Immunoassay Cellular Systems were used to select which Lumit secondary antibody pair performs the best. Lumit™ Anti-Mouse Ab-LgBiT and Anti-Rabbit Ab-SmBiT (set 1) was determined to be the optimal for pERK detection.
29. To reach an optimal signal, cell number should be determined experimentally because the signal is dependent on the cell type and signaling pathway. Generally, an optimal signal change can be obtained in a range of 10,000–80,000 cells per well of a 96-well plate. For suspension cells, double number of cells is usually seeded to perform Lumit Immunoassay immediately.

30. Remove or add the medium by contacting pipette tip to the wall of wells to avoid losing cells.
31. GF-AFC Substrate is a cell viability substrate for cell number normalization. This is a nonlytic, single-reagent-addition fluorescence assay component that measures the relative number of viable cells in a population. It is based on measurement of a conserved and constitutive protease activity within live cells that becomes inactive when cells die and therefore serves as a biomarker of cell viability. The substrate enters intact cells, where it is cleaved by the live-cell protease to generate a fluorescent signal proportional to the number of living cells. The GF-AFC substrate can be added to cells 30 min to 3 h before lysis. After lysis, the accumulated fluorescence is measured at the end of the Lumit Immunoassay along with luminescence (Fig. 6c, d). Fluorescence can also be read before the lysis step to check for cell viability.
32. Antibody incubation time can be determined experimentally depending on the antibody affinity.
33. When GF-AFC fluorogenic substrate is used to determine cell viability, the following can be used for data analysis:
    (a) Divide luminescence values by fluorescence values from the same well.
    (b) Calculate the average of entire fluorescence values.
    (c) Multiply each value from **step 33a** by the fluorescence value in **step 33b** to get normalized relative luminescence values.
    (d) Subtract the normalized relative luminescence values of "Lumit reagent" control from the normalized relative luminescence values of each of the treatments to obtain net luminescence values.
    (e) Plot the data.
34. Here, Lumit secondary antibodies are used to detect Fc-tagged proteins. However, other tags can be used with a pair of anti-tag antibodies (e.g., anti-His and anti-GST).

Acknowledgments

We thank James J. Cali for his critical reading of the manuscript and Julia Gilden for generating differentiated human T helper 2 (Th2) cells.

References

1. Miyai K (1991) Classification of immunoassay. In: Price CP, Newman DJ (eds) Principles and practice of immunoassay. Palgrave Macmillan, London, pp 246–264
2. Dixon AS, Schwinn MK, Hall MP, Zimmerman K, Otto P, Lubben TH, Butler BL, Binkowski BF, Machleidt T, Kirkland TA, Wood MG, Eggers CT, Encell LP, Wood KV (2016) NanoLuc complementation reporter optimized for accurate measurement of protein interactions in cells. ACS Chem Biol 11:400–408
3. Hall MP, Unch J, Binkowski BF, Valley MP, Butler BL, Wood MG, Otto P, Zimmerman K, Vidugiris G, Machleidt T, Robers MB, Benink HA, Eggers CT, Slater MR, Meisenheimer PL, Klaubert DH, Fan F, Encell LP, Wood KV (2012) Engineered luciferase reporter from a deep sea shrimp utilizing a novel imidazopyrazinone substrate. ACS Chem Biol 7:1848–1857
4. Bodle CR, Hayes MP, O'Brien JB, Roman DL (2017) Development of a bimolecular luminescence complementation assay for RGS: G protein interactions in cells. Anal Biochem 522: 10–17
5. Duellman SJ, Machleidt T, Cali JJ, Vidugiriene J (2017) Cell-based, bioluminescent assay for monitoring the interaction between PCSK9 and the LDL receptor. J Lipid Res 58:1722–1729
6. Norisada J, Fujimura K, Amaya F, Kohno H, Hirata Y, Oh-hashi K (2018) Application of NanoBiT for monitoring dimerization of the Null Hong Kong variant of α-1-antitrypsin, NHK, in living cells. Mol Biotechnol 60:539–549
7. Hwang B, Engel L, Goueli SA, Zegzouti H (2020) A homogeneous bioluminescent immunoassay to probe cellular signaling pathway regulation. Commun Biol 3:8
8. Nath N, Flemming R, Godat B, Urh M (2017) Development of NanoLuc bridging immunoassay for detection of anti-drug antibodies. J Immunol Methods 450:17–26
9. Turner MD, Nedjai B, Hurst T, Pennington DJ (2014) Cytokines and chemokines: at the crossroads of cell signalling and inflammatory disease. Biochim Biophys Acta Mol Cell Res 1843:2563–2582
10. Lumit immunoassay cellular systems technical manual. https://www.promega.com/Resources/Protocols/Technical%20Manuals/500/lumit-immunoassay-cellular-systems-protocol/. Accessed 8 Jan 2021

Chapter 16

ELISA-Based Biosensors

Robert S. Matson

Abstract

Enzyme-linked immunosorbent assay (ELISA) is by definition a biosensor. However, not all immunobiosensors involve the use of enzymes, while other biosensors incorporate ELISA as a key signaling component. In this chapter, we review the role of ELISA in signal amplification, integration with microfluidic systems, digital labeling, and electrochemical detection.

Key words Biosensor, Bio-recognition, Transducer, Analog, Digital, Microfluidics, Lab-on-a-chip (LOC), Limit of detection (LOD), Electrochemiluminescence (ECL), Enzyme-labeled fluorescence (ELF), Tyramide (TSA), Minimal detectable dose (MDD), Digital ELISA (dELISA), Digital droplet ELISA (ddELISA), Pre-equilibrium digital ELISA (PEdELISA), Liposome, Electrochemical, Working electrode, Potentiometric, Impedance, Amperometric, Ruthenium, Luminol, Quantum dots

1 Introduction

In a broader sense, enzyme-linked immunosorbent assay (ELISA) is a biosensor, that is, "a device which uses a living organism or biological molecules, especially enzymes or antibodies, to detect the presence of chemicals" (Oxford Languages). Bhalia et al. [1] further describe the essential features and components of the biosensor (*see* Fig. 1). First, within a defined environment (e.g., plate well + sample) an *analyte* must interact or be recognized by the *bioreceptor*. In ELISA, the antigen in the sample is recognized by the capture antibody immobilized to the well. To ascertain that this *bio-recognition* event took place, that is, the antibody has captured (sequestered) the antigen, a signal of some form needs to be generated. This is the job of the *transducer* that can detect a change within the environment. Thus, the secondary antibody–enzyme conjugate that binds to the sequestered antigen ultimately becomes the transducer when the enzyme converts substrate to a colored product. We then know that an analyte in the sample has been detected in this case by a change in color. Perhaps this should be considered a qualitative measure unless the color produced can be

Robert S. Matson (ed.), *ELISA: Methods and Protocols*, Methods in Molecular Biology, vol. 2612,
https://doi.org/10.1007/978-1-0716-2903-1_16,

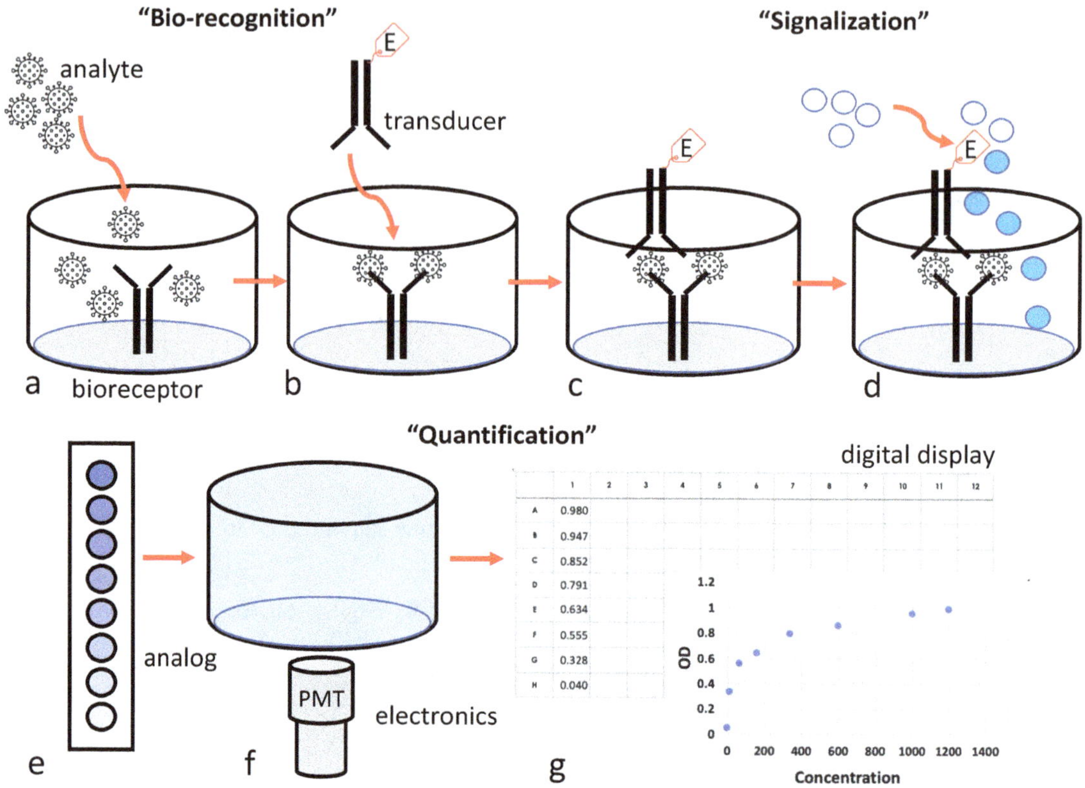

Fig. 1 The ELISA biosensor. A conventional microplate colorimetric ELISA is presented in terms of the essential elements of a biosensor described by Bhalia et al. [1]. Here, in the *bio-recognition phase* (**a**) the *analyte* [antigen (Ag)] is brought in contact with the *bioreceptor* [capture antibody (cAb)]. Upon binding of the analyte to the bioreceptor (Ag–cAb), the *transducer* [enzyme antibody conjugate (Enz-Ab)] is introduced (**b**) where it binds (**c**) to form the complex (cAb-Ag-Ab-Enz). The *signalization phase* (**d**) occurs when enzyme catalysis of substrate to colored product takes place. *Quantification* involves the detection of the colored (analog) signal (**e**) and (**f**) conversion by electronic means [e.g., photomultiplier tube reader (PMT)] to a digital signal display (**g**)

compared against a color chart of known values. Otherwise, to quantify the signal an electronic interface is often needed to convert from analog to digital (*signalization*) and the conditioned signal processed for display.

While an ELISA plate assay meets the definition of a biosensor, we often look to the lab-on-a-chip (LOC) as a more representative example. Here, we envision miniaturization, on-board controllers, detectors, microfluidics, valving, mixing, heating, and cooling with integration of these subsystems within a singular device.

2 Integration of ELISA with Microfluidics

Sun et al. [2] introduced the ELISA-LOC in which a miniature 96-well plate was addressed by a microfluidics system for detection of enterotoxins by ELISA. The plate wells (3 mm diameter) were designed to hold 13 µL of fluid. Each well had a 0.2-mm drain hole that interfaced with an outlet microfluidic drain system from which to draw vacuum. The plate cover that sealed the wells contained fluidic channels from which to deliver reagents to each of the 96 wells (single-node distribution) or to groups of 24 wells (four-node configuration) with distribution splitters. Samples were first loaded into wells, then the desired plate cover with fluidics placed over the wells and sealed. Reagents (wash buffer, enzyme conjugate, substrate) were loaded into fluidic loading chambers located on the plate top that could be pressurized via syringe for delivery. ELISA was designed with capture antibody bound to carbon nanotubes as the solid phase that absorbed to the well bottoms. Vacuum via syringe or peristaltic pump was applied to the outlet to move fluids out of the wells. ELISA comprised HRP-mediated electrochemiluminescence (ECL) in which the generated light was detected by a CCD camera mounted above the well plate. A limit of detection (LOD) ~0.1 ng/mL of enterotoxin was established. The authors note that this is comparable to standard ELISA for enterotoxin measured between 0.1 and 10 ng/mL.

Thaitrong et al. [3] adopted the Optimiser™ (Siloam Biosciences) microfluidic cartridge system in the detection of plant pathogens. The microfluidic system is a capillary spiral channel reaction chamber that fits into the well of a 96-well microplate. The microchannel device reduces the assay volume to about 5 µL permitting rapid concentration (~10 min/loading) of analyte from multiple loadings. This leads to an increase in sensitivity and is scalable when employed with liquid-handling robotics [4]. In their work, monoclonal capture antibodies to various plant pathogens were passively adsorbed to the device. All steps from blocking to signal generation were performed as with the steps generally employed for processing a conventional ELISA. The Optimiser™ ELISA (*see* Fig. 2), depending upon the immunoassay format, requires about 75–90 min from capture antibody loading and blocking to read [3, 5]. A conventional microplate-based format can take around 4.5 h if you include the coating and blocking steps. Kai et al. [5] reported sensitivity for an IL-4 cytokine assay in the range of 8–1000 fg/mL and an LOD ~2 fg/mL, while a standard ELISA produced a linear range of 1–125 pg/mL. Thaitrong et al. [3] found a 2- to 12.5-fold increase in sensitivity in the detection of plant pathogens when compared side-by-side with a standard ELISA. Moreover, the spiral capillary microfluidic device required a 20-fold lower sample volume while reducing the assay time by at least 50%.

Recently, Gosh et al. [6] described an automated (albeit, complex) microfluidic system incorporating spiral capillary reaction tubes capable of performing an autonomous ELISA. In this case, three independent spiral chambers were incorporated into the

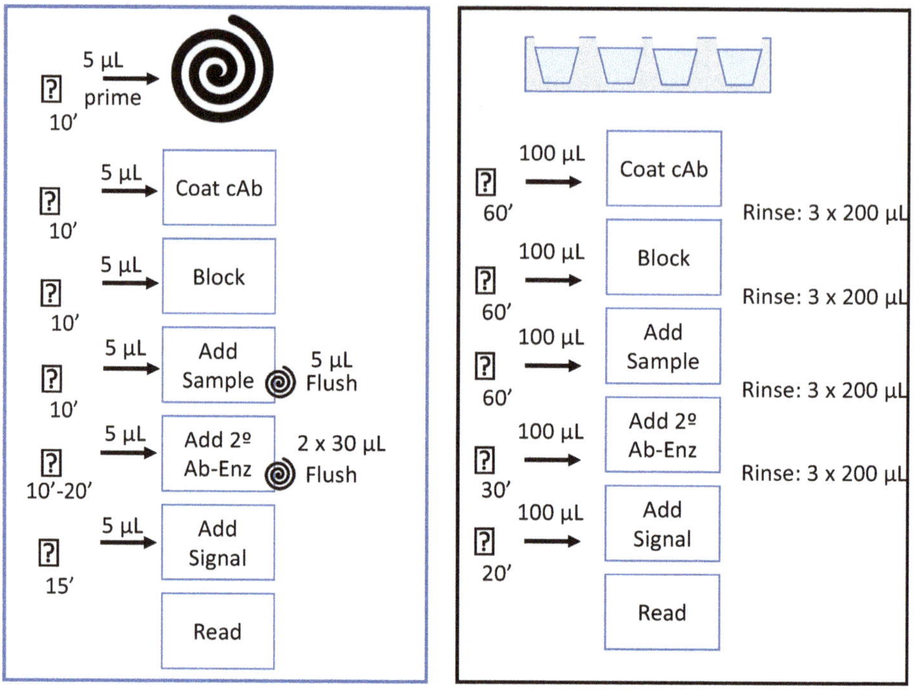

Fig. 2 Comparison of the Optimiser™ and standard ELISA processes. The Optimiser™ (Siloam Biosciences) is a spiral capillary microchannel device designed as a microplate well insert (**a**). Rapid processing times (10 min per step) and low-volume loading (5 μL) are reported to improve sensitivity and reduce assay time [3–5]. In this flowchart, the processing time from antibody coating to read for the device was approximately 75 min, while a conventional microplate ELISA (2B) takes about 4 h. The savings in sample and reagent consumption using the microfluidic device are evident

design consisting of the test chamber, positive control chamber, and negative control chamber. The micro-channel flow assay (MCFA) runs without external pumping based upon channel geometry and wetting properties. A smartphone is interfaced with the cartridge system and on-board ultrasensitive detector via a USB-OGT (On-the-Go) port. Thus, software associated with the smartphone is used to drive the optical device and process the captured images from the assay. An HRP-mediated chemiluminescent assay (ELISA-ECL) was employed as previously described. An important feature of the device was the capability of storing dry reagents on the platform, including secondary antibody HRP conjugate and a single component or one-step chemiluminescent substrate. Placement of hydrophilic regions within the capillary aids in pulling sample solution into the dried chemiluminescent substrate (detection reagent) for reconstitution and mixing. The sample is split into two paths. One path is directed toward the detection reagent; and the parallel path stream allows entry of the secondary antibody and sample into the chamber for binding to the immobilized capture antibody well in advance of the detection reagent. In this manner, the MCFA mimics the conventional steps used in an ELISA. An approximately 100-fold improvement in sensitivity was reported with the MCFA when compared with a commercial immunoassay for malaria.

3 Signal Enhancements: Enzymatic Amplification

3.1 Enzyme-Labeled Fluorescence (ELF)

The enzymatic signal output from ELISA is typically colorimetric or chemiluminescent. However, alkaline phosphatase catalysis of the ELF-97 alcohol substrate results in the formation of a highly fluorescent photostable precipitate with a Stokes shift at greater than 180 nm in wavelength. In previous works, we have demonstrated its use in DNA hybridization and multiplex immunoassays. For example, a well-based microarray was assessed for IL-4, IL-8, and IL-10 cytokines with ELF signal. For IL-8, a dynamic range 16–1000 pg/mL was achieved with a minimal detectable dose (MDD) at ≤ 1 pg/mL [7]. The drawback is that an ultraviolet light source is required for optimal excitation at 360 nm to achieve a 530 nm emission [8]. ELF is also useful in the detection of endogenous phosphatase activity and immunohistochemistry (HIC) labeling of cells and tissues [9].

3.2 Tyramide Signal Amplification (TSA)

A similar enzyme-mediated fluorescent signal amplification process, the tyramide signal amplification (TSA), has been widely adopted for tissue staining techniques such as IHC and in situ hybridization (ISH) as well as ELISA. The technology was first described by Bobrow et al. [10] as catalyzed reporter deposition (CARD) in which HRP catalyzed the surface deposition of biotin-phenols, followed by a secondary sequestering of the streptavidin-HRP by the surface-bound biotin-phenols. This was found to greatly amplify signal. A good overview of the technique as well as protocols for IHC applications is provided in an earlier edition [11].

Stack et al. [12] describe the strategy for multiplex IHC based upon the use of different fluorescently labeled tyramides for sequential staining of tissue. Aydin [13] utilized TSA in a magnetic bead-based immunoassay to detect *Escherichia coli* O157:H7 in milk and ground beef tainted with this Shiga toxin-producing bacteria.

4 Digital ELISA

4.1 dELISA

"Counting molecules is the most accurate method for measuring the concentration of molecules in solution" [14]. This is the underlying theorem for single-molecule detection that has led to the rise of digital ELISA (dELISA). A rudimentary description is provided

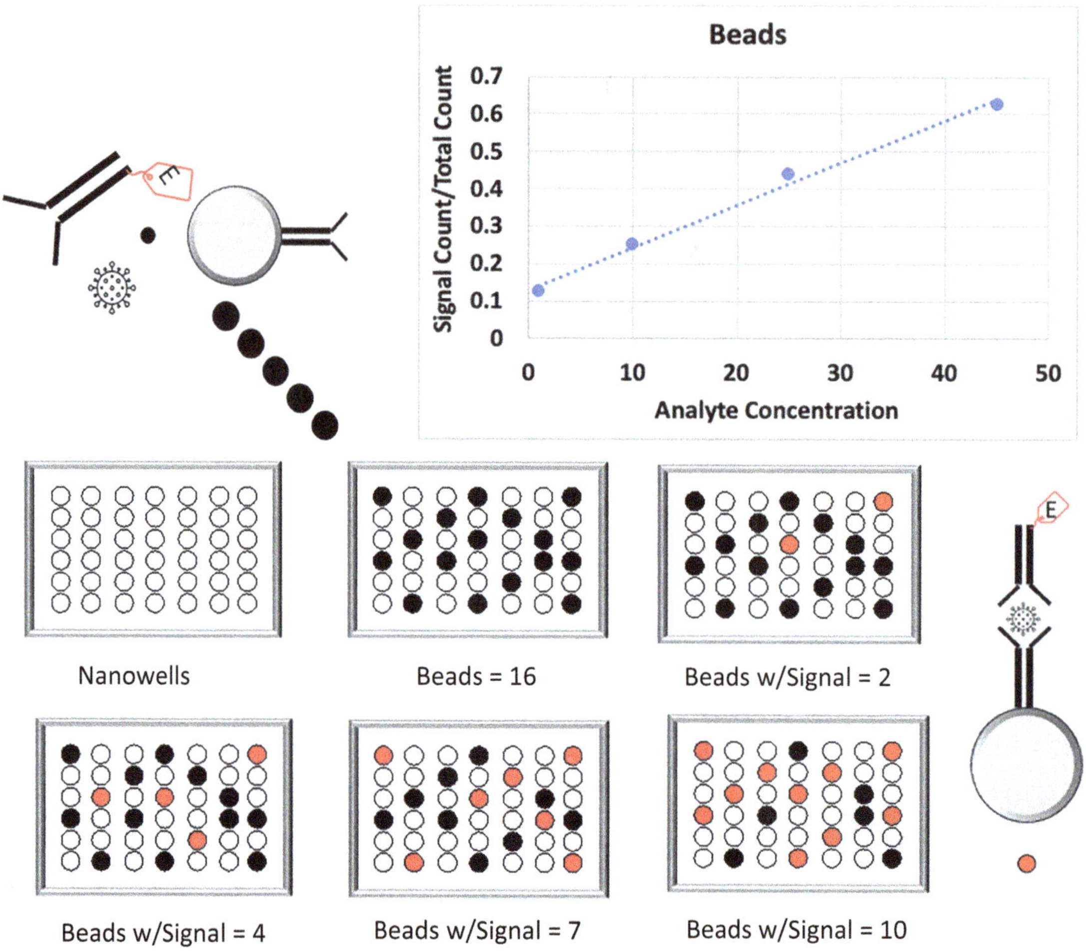

Fig. 3 Counting beads. The concept of digital ELIISA is based upon the ability to count nanobeads in nanowells. Each nanowell can accept only a single nanobead. Each nanobead contains capture antibody. Owing to the size of the bead, the number of tethered antibodies is limited due to surface area. As a result, an enzyme amplification is required to accurately count immunopositive beads

in Fig. 3. The platform originated from the work of Rissin and Walt [15], who tethered streptavidin β-galactosidase to biotin-bound beads. The enzyme beads (~2.7 μm diameter) were then dispersed into nanowells (~50 femtoliter volume) of a modified optical fiber. Each nanowell could hold only a single bead. The concentration of the enzyme was set to statistically allow either single binding or no binding to a bead. Thus, fluorescent signal from labeled substrate in a nanowell provided a single count (1) and the absence of signal from a bead provide (0) count. The input enzyme concentration was then varied over a specific input range and applied to different optical fibers. A total of 5000 nanowells were counted per fiber. Those wells counted as enzyme active (log % active wells) correlated ($r^2 = 0.9907$) with the input concentration of the enzyme (log moles of target). The dELISA platform was subsequently commercialized by Quanterix (Cambridge, MA) under the product trade name Single Molecule Arrays (SiMoA). Studies from the Quanterix group [16] demonstrated the ultrasensitive detection of prostate-specific antigen (PSA) and TNF-α at LOD of 6 fg/mL (200 aM) and 10 fg/mL (600 aM), respectively. In contrast, the authors reported at the time on detectable levels from commercial ELISA platforms: PSA LOD at 0.1 ng/mL (3 pM) and TNF-α at 0.34 pg/mL (21 fM).

4.2 ddELISA

Recently, Cohen et al. [17] introduced droplet digital ELISA (ddELISA), which combines dELISA with droplet microfluidics (such as utilized in ddPCR). The platform's intent is to address the detection of very-low-abundant biomarkers potentially useful in early diagnosis of certain cancers. Thus, capture antibody was immobilized on 2.7-μm-diameter beads and incubated with sample (~100,000 beads/sample), followed by addition of biotinylated secondary antibody and streptavidin β-galactosidase reagents in a 96-well microplate. After 2-h incubation, the beads were resuspended in fluorescent substrate and injected into oil droplets (1.4 pL) using a microfluidic droplet generator system. Subsequently, the bead droplets were transferred to an imaging chamber to form a monolayer. The droplets were imaged in brightfield for mapping bead locations and fluorescence to identify beads with positive signal. The authors compared ddELISA results with dELISA (Simoa System) for cytokines IFNγ and IL-2. For IFNγ, LOD 30 aM versus 350 aM, and IL-2, LOD 20 aM versus 550 aM, respectively. Thus, for these analytes a significant increase in sensitivity was realized using the ddELISA platform.

4.3 TSA-dELISA

Akama et al. [18] introduced a droplet-free digital ELISA (dELISA) technique employing TSA for the analysis of hepatitis B surface antigen (HBsAg), which is a biomarker for hepatitis B virus (HBV). The tyramide-labeled beads were quantifiable by flow cytometry, and the results were comparable ($r^2 = 0.98$) with a droplet-based dELISA but at lower detection efficiency based upon the slope. Nevertheless, the authors report an LOD at 139 aM, which is a 20-fold improvement in sensitivity relative to a standard ELISA.

Maley et al. [19] sought to capture TSA-labeled beads onto a fibrin hydrogel for display. The intended effort was aimed at providing a CARD-dELISA device for point-of-care (POC) applications. A microscope brightfield image profile of the encapsulated beads served to map the location of individual beads. Fluorescent images from those beads capturing analyte (labeled with tyramide-Alexa 488) were overlayed with the corresponding brightfield image for counting. IL-6 cytokine was evaluated using the CARD-dELISA, which achieved an LOD of 1.36 fM.

Finally, Fu et al. [20] coupled TSA with surface-enhanced Raman spectroscopy (SERS) detection. In this case, gold nanoparticles were labeled with a SERS probe (4-mercaptobenzoic acid) and subsequently conjugated with tyramine. A gold-coated silicon wafer was loaded with antimouse IgG capture antibody. A secondary antibody (antimouse IgG-HRP conjugate) was added to sandwich mouse IgG antigen. The SERS-tryamine particles were added and the TSA reaction triggered with H_2O_2. Following aggregation of particles on the wafer, the surface was interrogated by laser excitation (633 nm) and the resultant Raman signal evaluated. The linear dose–response (0.01–100 ng/mL) was measured from the intensity band observed at 1583 cm^{-1}. An LOD of 0.01 ng/mL mouse IgG was reported.

4.4 Pre-equilibrium Digital ELISA (PEdELISA)

To bring an ultrasensitive multiplex ELISA platform to the bedside for real-time monitoring of biomarkers during acute critical care, Song et al. [21] introduced a rapid quench of bead-analyte binding during dELISA. That is, the analyte capture was not allowed the time to reach completion (1–2 h), but rather capture was halted under "pre-equilibration" conditions of 15 s–300 s. Hence, the new technique of PEdELISA. Essentially, in a two-step immunoassay they found that these very short incubation times were sufficient to achieve a linear dynamic range from femtomolar (fM) to nanomolar (nM) levels of cytokines. For example, the authors report upon achieving for IL-6 an LOD = 25.9 pg/mL within 15 s at a linear dynamic range up to 10 ng/mL. For a full cytokine 10-plex, the LODs at 300 s ranged from 0.2 to 4.3 pg/mL. This permits monitoring of a patient's blood profile for hyper-immunoreactivity associated with a cytokine storm during chimeric antigen receptor T-cell (CAR-T therapy). In this study, levels of cytokines for 10 patients experiencing moderate to severe grades of cytokine release syndrome (CRS) were monitored during their therapy.

5 Other Immunosensor Platforms

5.1 Temperature-Responsive Liposome-Linked Immunosorbent Assay (TLip-ELISA)

Hu et al. [22] describe the use of liposomes loaded with fluorescent dye to attenuate signal. A highly fluorescent squaraine dye, SQR22, incorporated into liposomes undergoes autoquenching due to a phase transition change in the liposome upon cooling. This is reversible with an increase in temperature. Thus, the dye-loaded liposome can serve as an on/off signal switch [23]. The TLip-ELISA was designed with liposome (TLip) incorporated with biotinylated phospholipid and SQR22 as the reporter (biotin-TLip). For the determination of PSA, an anti-PSA capture antibody was coated down onto a microplate well. PSA protein was first incubated, followed by the addition of biotinylated anti-PSA reporter antibody. Next, streptavidin was added to enable bridging between the biotinylated reporter antibody and biotin-TLip, which was added last. The microplate was then heated from 27 °C to 80 °C to permit the phase transition, which resulted in an increase in fluorescent signal. A linear dynamic range for PSA was established at 1 (fg)/mL to 0.01 fg/mL or 10 attogram (ag)/mL. The LOD was determined to be 27.6 ag/mL.

5.2 Electrochemical Signal

Wujcik et al. [24], in their review, described that the fundamental characteristic associated with "electrochemical antibody nanosensors" is measuring the conformational change that occurs during biorecognition between antibody and antigen. In this instance, transduction via the generation of an electric field signal. A working electrode provides the surface for immobilization of the capture antibody and site of the electrochemical transduction, while the counter electrode completes or accepts the generated electrolyte. The reference electrode maintains the potential (mV) on the working electrode. We can then measure any difference in electric signal by *amperometry* (change in current at constant potential), *potentiometry* (change in potential at approximately zero current), or *impedimetric* technique (change in resistance to current flow or capacitance).

Impedance-Based Immunosensors

Biosensors based upon impedance do not require enzyme catalytic activity. Therefore, impedance is not measured with ELISA and is considered a label-free detection technique. It is generally observed that the impedance increases as the complexity increases on the sensing electrode, for example, the binding of antigen to antibody on the electrode surface may mask or prevent electrolyte contact with the electrode, i.e., building up resistance to current flow. It is also possible in some cases that the analyte binding leads to a distortion of the underlying polymer coating that exposes the electrode, thereby increasing current flow [25].

Electrochemical impedance spectroscopy (EIS) is a very useful technique for measuring binding kinetics of ions, small molecules, DNA, and proteins [26]. Microfluidic impedance flow cytometry is used to interrogate single cells measuring cell size, membrane capacitance, and subpopulation differences [27].

In an early study, Teeparuksapun et al. [28] describe a flow injection capacitance immunosensor for the detection of HIV capsid protein p24. Gold nanoparticles (AuNP) with adsorbed anti-p24 capture antibody (cAb) were bound to the working (Au) electrode. The p24 antigen (analyte) was injected and subsequently captured on the Au-AuNP-cAb electrode. Change in capacitance was measured with increased injection loading concentration of the p24 antigen. An LOD of 7.9×10^{-8} pg/mL (79 ag/mL) was calculated. This study confirmed the very high sensitivity achievable by an impedance immunoassay.

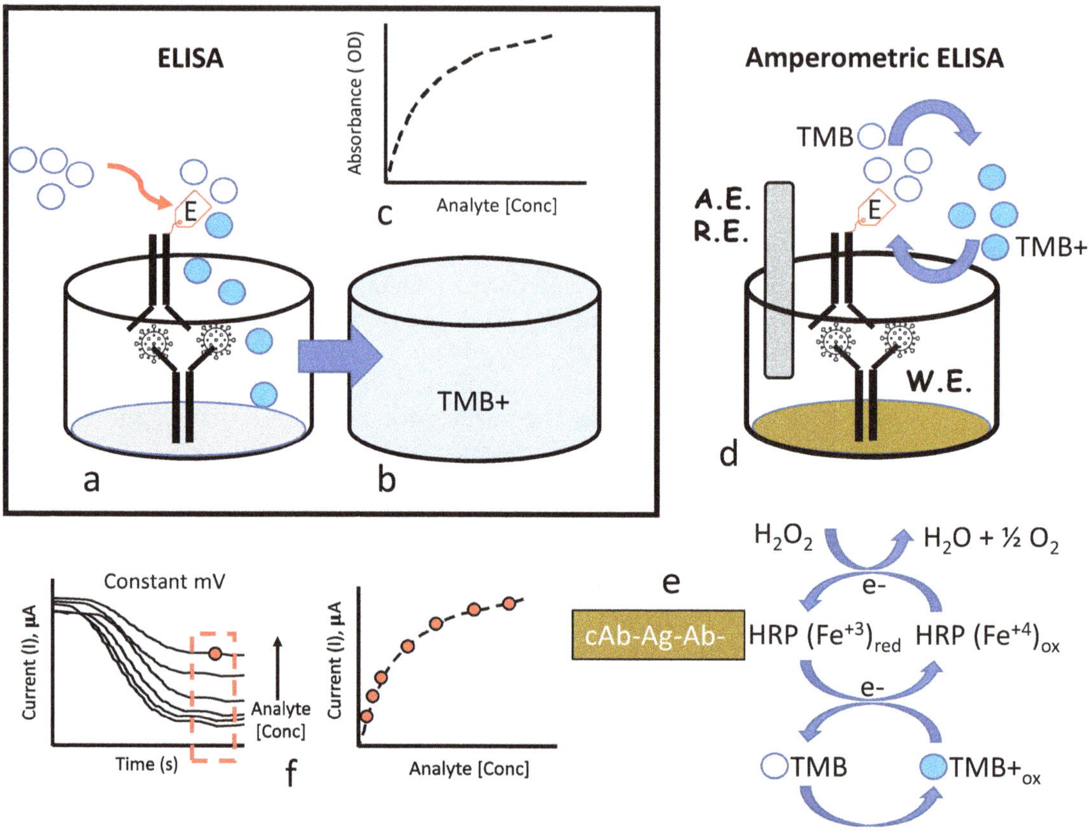

Fig. 4 Colorimetric ELISA versus amperometric ELISA. For the conventional HRP/TMB colorimetric ELISA, oxidation of TMB (**a**) leads to a blue color (**b**), which is proportionate to the input concentration of analyte (**c**). In the case of the amperometric ELISA, the TMB is oxidized, then reduced (**e**) at the working electrode (**d**). The current is measured from the redox reaction over time at a constant potential (**f**). Working electrode (W.E.), auxiliary electrode (A.E.), reference electrode (R.E.), oxidized TMB (TMB+), and reduced (TMB)

Amperometric ELISA

"The most common amperometric immunosensor is the sandwich ELISA with electrochemical detection" [29]. Essentially, an enzyme such as HRP is used to create a redox potential by electron transfer from the bound immunocomplex to the working electrode surface (*see* Fig. 4). An electric potential is set between the working electrode (W.E.) and the auxiliary electrode (A.E.), resulting in a constant potential between the W.E. and reference electrode (R.E.). The change in current over time is measured. The difference between the initial and final levels or the rate of current increase is used for the determination of analyte concentration.

Benkert et al. [30] developed a creatinine immunosensor based upon a competitive ELISA format. Creatinine is a clinical marker associated with kidney function. To construct the sensor, creatinine (antigen) was immobilized onto a cellulosic membrane. The creatinine membrane was then placed next to the working electrode set at an applied potential of +600 mV versus Ag/AgCl. Anti-creatinine monoclonal antibody served as the competitive antibody, while glucose oxidase antimouse antibody conjugate served as the reporter. Sample containing creatinine analyte was mixed with a cocktail of competitor and reporter antibodies and applied to the sensor. After a rinse, glucose was added and the increase in current measured due to the glucose oxidase-mediated redox of hydrogen peroxide: α-D-Glucose + $O_2 \rightarrow$ D-Glucono-1,5-lactone + H_2O_2. An LOD of 4.5 ng/mL creatinine (40 nM) was achieved. The measured range was reported at 0.01–10 μg/mL (0.09–90 μM). The current normal serum creatine level is reported to be in the range of 0.6–1.3 mg/dL corresponding to about 50–115 μM.

Martínez-García et al. [31] describe a screen-printed carbon electrode (SPCE) system for amperometric ELISA detection of amylin, a biomarker for type 2 diabetes and obesity. The SPCE working electrode was first coated by electro-polymerization of poly (pyrrole propionic acid) (pPPA), which provided carboxyl groups for attachment of an anti-amylin capture antibody. The competitive ELISA was set up using biotinylated amylin versus native amylin (analyte). Streptavidin-HRP was added for redox of hydroquinone (HQ)/H_2O_2 at an applied potential of −200 mV. Here, HQ is oxidized to benzoquinone (BQ) in the HRP-mediated catalysis of hydrogen peroxide: HQ + $H_2O_2 \rightarrow$ BQ + H_2O. A linear dynamic range of 1 fg/mL to 50 pg/mL was reported with an LOD of 0.92 fg/mL. Amylin spiked into serum (1.5–8 pg/mL) and urine (0.75–4 pg/mL) showed recoveries within 83–107%.

ECL Immunosensors

ECL basically utilizes electrode-driven electron-transfer reactions to excite various chemical species, leading to the release of light. Most current systems involve a working electrode, a light emitter, and a co-reactant species that can mediate electron transfers, leading to the generation of free radicals. The radicals decompose and excite the emitter, leading to the release of light. For a sandwich immunoassay, the capture antibody is immobilized onto or near the electrode surface. The reporter antibody is typically conjugated with ECL emitter.

There are three types of co-reactants the ECL emitter reagent systems is based upon metal ions (ruthenium, Ru^{+2}, or iridium, Ir^{+3}), luminol, or the use of nanoparticles (quantum dots). The *ruthenium systems* usually consists of tris(2,2′-bipyridyl) ruthenium (II) or $Ru(byp)_3^{+2}$ caged complex as the emitter and tripropylamine (TPA) as the co-reactant [32]. Reactivity is influenced by the choice of electrodes [33]. In other work, Zanut [34] explores the mechanism and role of the co-reactants for the $Ru(byp)_3^{+2}$ system that may lead to improvements in signal enhancement for ultrasensitive immunoassays.

The *luminol systems* comprise luminol as the emitter and H_2O_2 as the co-reactant. Hence, redox enzymes such as HRP can participate in peroxide generation. The *nanoparticle (NP) systems* are rapidly evolving as new materials become available such as quantum dots (semiconductor nanocrystals), bimetallic nanoparticles, graphene, carbon nanotubes, fullerenes, and others. The advantages of nanomaterials for immunoassay include biocompatibility and enhanced surface area. These features improve electrode efficiency, increased surface density for antibody immobilization, and reduced detection times [35].

Commercial ECL systems have been reviewed by Y. Zhang et al. [36] including their linear dynamic ranges and LOD comparisons for the quantification of various drugs and metabolites of clinical interest.

References

1. Bhalla N, Pawan P, Formisano N, Estrela P (2016) Introduction to biosensors. Essays Biochem 60:1–8. https://doi.org/10.1042/EBC20150001
2. Sun S, Yang M, Kostov Y, Rasooly A (2010) ELISA-LOC: lab-on-a-chip for enzyme-linked immunodetection. Lab Chip 16:2093–2100. https://doi.org/10.1039/C003994B
3. Thaitrong N, Charlermroj R, Himananto O et al (2013) Implementation of microfluidic sandwich ELISA for superior detection of plant pathogens. PLoS One 8(12):e83231. https://doi.org/10.1371/journal.pone.0083231
4. Tecan (2012) Application note: automating ELISAs on Tecan's Freedom EVO® using Optimiser™ technology from Siloam Biosciences. Tecan, Männedorf
5. Kai J, Santiago N, Puntambekar A, et al (2011) The next generation microplate using power of microfluidics for femtogram/ml level sensitivity. 15th international conference on miniaturized systems for chemistry and life sciences, 2–6 October 2011, Seattle, WA, USA
6. Ghosh S, Aggarwal K, Vinitha TU et al (2020) A new microchannel capillary flow assay (MCFA) platform with lyophilized chemiluminescence reagents for a smartphone based POCT detecting malaria. Microsyst Nanoeng 6:5. https://doi.org/10.1038/s41378-019-0108-8
7. Matson RS (2013) Chapter 6: Protein microarray applications. In: Matson RS (ed) Applying genomic and proteomic microarray technology in drug discover, 2nd edn. CRC Press, Baco Raton, pp 224–225
8. Paragas VB, Zhang Y-Z, Haugland RP, Singer VL (1997) The ELF-97 alkaline phosphatase substrate provides a bright, photostable, fluorescent signal amplification method for FISH. J Histochem Cytochem 45(3):345–357
9. Johnson I, Spence M, eds (2010) Chapter 6: Ultrasensitive detection technology, Section 6.3 phosphatase-mediated signal amplification techniques. In: The molecular probes handbook. A guide to fluorescent probes and labeling technologies, 11th edn. Life Technologies, Carlsbad
10. Bobrow MN, Harris TD, Shaughnessy KJ, Litt GJ (1989) Catalyzed reporter deposition, a novel method of signal amplification to immunoassays. J Immunol Methods 125:279–285
11. Faget L, Hanasko TS (2015) Chapter 16: Tyramide amplification for immunofluorescence enhancement. In: Hanasko R (ed) ELISA methods and protocols, Methods in molecular biology 1318, Springer protocols. Human Press, New York, pp 161–172
12. Stack EC, Wang C, Roman KA, Hoyt CC (2014) Multiplexed immunohistochemistry, imaging, and quantitation: a review, with an assessment of tyramide signal amplification, multispectral imaging, and multiplex analysis. Methods 70:46–58. https://doi.org/10.1016/j.ymeth.2014.08.016
13. Aydin M, Herzig GPD, Jeong KC et al (2014) Rapid and sensitive detection of Escherichia coli O157:H7 in milk and ground beef using magnetic bead–based immunoassay coupled with tyramide signal amplification. J Food Prot 77(1):100–105. https://doi.org/10.4315/0362-028X.JFP-13-274

14. Walt DR (2013) Optical methods for single molecule detection and analysis. Anal Chem 85:1258–1263. https://doi.org/10.1021/ac3027178
15. Rissin DM, Walt DR (2006) Digital readout of target binding with attomole detection limits via enzyme amplification in femtoliter arrays. J Am Chem Soc 128:6286–6287. https://doi.org/10.1021/ja058425e
16. Rissin DM, Kan CW, Campbell TG et al (2010) Single-molecule enzyme-linked immunosorbent assay detects serum proteins at subfemtomolar concentrations. Nat Biotechnol 28(6): 595–599. https://doi.org/10.1038/nbt.1641
17. Cohen L, Cui N, Cai Y et al (2020) Single molecule protein detection with attomolar sensitivity using droplet digital enzyme-linked immunosorbent assay. ACS Nano 14:9491–9501. https://doi.org/10.1021/acsnano.0c023789491
18. Akama K, Shirai K, Suzuki S (2016) Droplet-free digital enzyme-linked immunosorbent assay based on a tyramide signal amplification system. Anal Chem 88:7123–7129. https://doi.org/10.1021/acs.analchem.6b01148
19. Maley AM, Garden PM, Walt DR (2020) Simplified digital enzyme-linked immunosorbent assay using tyramide signal amplification and fibrin hydrogels. ACS Sens 5:3037–3042. https://doi.org/10.1021/acssensors.0c01661
20. Fu C, Jin S, Shi W et al (2018) Catalyzed deposition of signal reporter for highly sensitive surface-enhanced Raman spectroscopy immunoassay based on tyramine signal amplification strategy. Anal Chem 90:13159–13162. https://doi.org/10.1021/acs.analchem.8b02419
21. Song L, Lachno DR, Hanlon D et al (2021) A digital enzyme-linked immunosorbent assay for ultrasensitive measurement of amyloid-β 1–42 peptide in human plasma with utility for studies of Alzheimer's disease therapeutics. Alzheimers Res Ther 8(58):1–15. https://doi.org/10.1186/s13195-016-0225-7
22. Hu R, Sou K, Takeoka S (2020) A rapid and highly sensitive biomarker detection platform based on a temperature-responsive liposome-linked immunosorbent assay. Sci Rep 10(18086):1–11. https://doi.org/10.1038/s41598-020-75011-x
23. Sou K, Chan LY, Arai S, Lee C-LK (2019) Highly cooperative fluorescence switching of self-assembled squaraine dye at tunable threshold temperatures using thermosensitive nanovesicles for optical sensing and imaging. Sci Rep 9(17991):1–12. https://doi.org/10.1038/s41598-019-54418-1
24. Wujcik EK, Wei H, Zhang X et al (2014) Antibody nanosensors: a detailed review. RSC Adv 4:43725–43745
25. Leva-Bueno J, Peyman SA, Millner PA (2020) A review on impedimetric immunosensors for pathogen and biomarker detection. Med Microbiol Immunol 209:343–362. https://doi.org/10.1007/s00430-020-00668-0
26. Mollarasouli F, Kurbanoglu S, Ozkan SA (2019) The role of electrochemical immunosensors in clinical analysis. Biosensors 9(86): 1–19. https://doi.org/10.3390/bios9030086
27. Kim S, Song H, Ahn H et al (2021) A review of advanced impedance biosensors with microfluidic chips for single-cell analysis. Biosensors 11(412):1–14. https://doi.org/10.3390/bios11110412
28. Teeparuksapun K, Hedstrom M, Wong EY et al (2010) Ultrasensitive detection of HIV-1 p24 antigen using nanofunctionalized surfaces in a capacitive immunosensor. Anal Chem 82(20): 8406–8411
29. Contreras-Naranjo JE, Oscar Aguilar O (2019) Suppressing non-specific binding of proteins onto electrode surfaces in the development of electrochemical immunosensors. Biosensors 9(15):1–23. https://doi.org/10.3390/bios9010015. www.mdpi.com/journal/biosensors
30. Benkert A, Scheller F, Schossler W et al (2000) Development of a creatinine ELISA and an amperometric antibody-based creatinine sensor with a detection limit in the nanomolar range. Anal Chem 72(5):916–921
31. Martínez-García G, Sánchez-Tirado E, González-Cortés A et al (2018) Amperometric immunoassay for the obesity biomarker amylin using a screen-printed carbon electrode functionalized with an electropolymerized carboxylated polypyrrole. Microchim Acta 185(323): 1–8. https://doi.org/10.1007/s00604-018-2863-x
32. Miao W, Choi J-P, Bard AJ (2002) Electrogenerated chemiluminescence 69: the tris(2,2′--bipyridine) ruthenium (II), (Ru(bpy)32+)/Tri-n-propylamine (TPrA) system revisited-a new route involving TPrA cation radicals. J Am Chem Soc 124:14478–14485. https://doi.org/10.1021/ja027532v
33. Parveena S, Chena Y, Yuana Y et al (2021) Electrochemiluminescence of [Ru(bpy)3]2+/tripropylamine at glassy carbon, platinum, and palladium electrodes. Sens Actuators Rep 3(100062):1–6

34. Zanut A, Fiorani A, Canola S et al (2020) Insights into the mechanism of coreactant electrochemiluminescence facilitating enhanced bioanalytical performance. Nat Commun 11(2668):1–9. https://doi.org/10.1038/s41467-020-16476-2
35. Zhang Z, Cong Y, Huang Y, Du X (2019) Nanomaterials-based electrochemical immunosensors. Micromachines 10(397): 1–19. https://doi.org/10.3390/mi10060397
36. Zhang Y, Zhang R, Yang X et al (2019) Recent advances in electrogenerated chemiluminescence biosensing methods for pharmaceuticals. J Pharm Anal 9:9–19

Index

Robert S. Matson (ed.), *ELISA: Methods and Protocols*, Methods in Molecular Biology, vol. 2612, https://doi.org/10.1007/978-1-0716-2903-1,

T

U

V

W

X

Z

www.ingramcontent.com/pod-product-compliance
Lightning Source LLC
Chambersburg PA
CBHW081756061025
33644CB00038B/894
* 9 7 8 1 0 7 1 6 2 9 0 2 4 *